ENVIRONMENTAL SCIENCE, ENGINEERING AND TECHNOLOGY

AIR POLLUTION EMISSIONS

ENVIRONMENTAL SCIENCE, ENGINEERING AND TECHNOLOGY

Additional books in this series can be found on Nova's website under the Series tab.

Additional E-books in this series can be found on Nova's website under the E-book tab.

ENVIRONMENTAL SCIENCE, ENGINEERING AND TECHNOLOGY

AIR POLLUTION EMISSIONS

DANIEL G. VASILIEV
AND
ROBERT A. KASK
EDITORS

Nova Science Publishers, Inc.
New York

For permission to use material from this book please contact us:
Telephone 631-231-7269; Fax 631-231-8175
Web Site: http://www.novapublishers.com

Library of Congress Cataloging-in-Publication Data

Air pollution emissions / editors, Daniel G. Vasiliev and Robert A. Kask.
p. cm.
Includes index.
ISBN 978-1-62100-453-0 (hardcover)
1. Air pollution. I. Vasiliev, Daniel G. II. Kask, Robert A.
TD883.A47815 2011
363.739'2--dc23

2011034486

Published by Nova Science Publishers, Inc. † New York

CONTENTS

PREFACE

In this book, the authors present topical research in the study of air pollution emissions. Topics discussed include opportunities for sustainable energy development and applications in building environments and air pollution controls; hydrocarbon air pollution laser monitoring; inventory and behaviour of air pollutants in an urban area of NW Spain and spatial modeling of air pollution based on traffic emissions.

Chapter 1 - People to rely upon oil for primary energy and this for a few more decades. Other conventional sources may be more enduring, but are not without serious disadvantages. The renewable energy resources are particularly suited for the provision of rural power supplies and a major advantage is that equipment such as flat plate solar driers, wind machines, etc., can be constructed using local resources and without the advantage results from the feasibility of local maintenance and the general encouragement such local manufacture gives to the build up of small-scale rural based industry. This chapter comprises a comprehensive review of energy sources, the environment and sustainable development. It includes the renewable energy technologies, energy efficiency systems, energy conservation scenarios, energy savings in greenhouses environment and other mitigation measures necessary to reduce climate change. This theme gives some examples of small-scale energy converters, nevertheless it should be noted that small conventional, i.e., engines are currently the major source of power in rural areas and will continue to be so for a long time to come. There is a need for some further development to suit local conditions, to minimise spares holdings, to maximise interchangeability both of engine parts and of the engine application. Emphasis should be placed on full local manufacture. It is concluded that renewable environmentally friendly energy must be encouraged, promoted, implemented and demonstrated by full-scale plan especially for use in remote rural areas.

Chapter 2 - The hydrocarbons molecules of the different composition are the most pollutants in the atmosphere under the oil - gas converting or distributing enterprises. These molecules in atmospheric air monitoring can be realized by the laser remote sensing systems using. This allows to measure the pollutants concentration at the large sensing distances up to tenth of kilometers with the high repetition rate or time resolution. The experimental studies and computer simulation results were stated that the differential absorption and scattering lidar and Raman lidar have the preference possibilities in this problem solving as have been seen from the authors' previous papers.

Chapter 3 - This chapter reports the PM10 concentration measured at LUT and BON, two urban air quality monitoring stations located in the Saint-Pierre city, to the South of Réunion

Island. These two atmospheric stations are found close to each other (distant by 1.5 km) and to the sea (respectively at 500 and 200 m from it). The PM10 concentrations measured continuously with automatic analysers from August 2007 to August 2008 at these two sites were studied in conjunction with meteorological data (winds and rainfall) also recorded at Saint Pierre and the swell data recorded at Saint Pierre and other parts of the island.

Chapter 4 - Directives for evaluating the ambient air quality in the EU have been adopted in the last years as a part of the new strategies for pollution prevention and control. Such strategies are based in the control of emissions and measurement of gaseous and particulate pollutants in the air, in industrial, urban and rural areas. Inventories of industrial emissions and concentrations measured in control stations of NO_2, NO, O_3, SO_2, CO, CO_2 and particulate matter have been monitored in urban areas of the EU. As a result, the knowledge of dispersion, interaction and general behaviour of such pollutants have increased. Both the dispersion of pollutants, and daily, weekly, sesonal and annual cycles caused by the emissions have been studied in different areas or the World by using such control stations. In the NW of Spain, available data from emissions and measurements of pollutants in the stations have provided valuable information about the behaviour of the pollutants in the urban area of A Coruña city. In this area the behaviour of pollutants is marked by the seaside climate. The data compiled in the year 2007 are presented here. They have allowed understand the main sources causing dangerous concentrations of pollutants, to use such information for future urban planning. Results of such observations indicate that near sources of air pollutants can be responsible of mean and peak concentrations. Also, the trend of some pollutants can be related to climatic factors.

Chapter 5 - Volatile organic compounds (VOCs) constitute a heterogeneous group of substances with different physical and chemical behaviours but with the common feature of having high vapour pressures. When these compounds reach the atmosphere they undergo fast reactions with ozone, OH and nitrate radicals and, as a consequence of these reactions, it is produced a great variety of pollutants. These secondary pollutants include ground-level ozone, peroxyacetyl nitrate, carbonyl compounds or organic acids. Additionally, the generated products also contribute to the formation of secondary organic aerosols (SOA) in the atmosphere. Apart from these harmful effects on the environment, VOCs also produce acute and chronic effects on human health that may range from simple sensory irritations to severe behavioural, neurotoxic and genotoxic damages.

In the monitoring programs the most suitable analytical method has to be selected taking into account the nature of the analytes of interest, the composition of the matrix and the concentrations of the target analytes in such matrix. The assurance of a representative sampling, the correct manipulation and storage of the sample and the precise and accurate determination of the target analytes are key parameters in the environmental monitoring programmes.

The aim of this chapter is to give an overview of the methodologies used for the sampling and analysis of VOCs in the atmosphere summarizing their main advantages and limitations. The methods discussed here include the traditional whole air sampling techniques, the cryogenic trapping and the adsorptive enrichment on solid adsorbents in the sampling stage and the use of gas chromatography combined with different detection systems for the separation and detection of the target analytes. The possible applications of several newer analytical methods for the analysis of VOCs in air are also discussed.

Chapter 6 - Road traffic becomes a dominant source of air pollution in urban areas. The emissions of inorganic compounds and volatile organic compounds caused by motor vehicles have increased trends, particularly in urban areas, which together with other emitted compounds can cause human health problems in a long-time period. Procedures for estimating of road traffic emissions in the USA and in the European region are based on guidelines and recommendations. In order to implement the guidelines, decision-making tools are necessary to provide information in an easy understandable form. Thus, the spatial information system is needed to manage spatio-temporal data, provide analysis, solve numerical models and visualize the results. Generally, spatial data include street networks and monitoring networks. Temporal data are represented by changes of traffic intensity and by time series of measured pollutants. Geographic information systems (GISs) extended by environmental modeling tools offer advance estimates of traffic emissions improved by the digital terrain data and by the wind flow effects. The estimates of air pollution in the surrounded areas can be based on spatial interpolation. In a wide range of spatial interpolations, the deterministic techniques or the geostatistical methods are used in dependence on the input data and the results provided by the exploratory spatial data analysis (ESDA). Each spatial interpolation attached to the map layer defines concentration levels at a specified time period. The time series of map layers can demonstrate the variability of the air pollution distribution, which brings a new insight into this research.

In addition to mapping of the urban air pollution in dependence on the traffic intensity and the measured concentrations, numerical modeling and simulation in wind tunnels are presented as important tools for exploration of the dispersion and the transport of pollutants caused by wind flows and other effects. A number of software tools based on Gaussian dispersion principles are mentioned in the framework of US EPA guidelines.

As a case study, spatial modeling of air pollution is focused on the urban area of the city of Prague, because a number of motor vehicles registered on the Prague territory is growing. In addition to the registered vehicles, the specific phenomenon in the central Europe is represented by the abrupt increase in traffic of trucks. In spite of new methods for reduction of emissions from motor vehicles, urban development and successive reconstructions of existing roads cause other emissions. Thus, the street network also becomes a set of line sources of stirred up suspended particulates generated by the passing vehicles. These dust emissions, so-called secondary dust, are estimated and partially validated by measurements of the fraction PM_{10}. As an example, estimates of nitrogen oxides are carried out by spatial interpolation based on data from the surface monitoring network and the DIAL-LIDAR measurements.

Integration of spatio-temporal data together with environmental modeling tools brings new possibilities to compare spatial interpolations created for individual compounds at the high temporal and spatial resolution. In the framework of the GIS, mapping of air pollution and human exposures in streets, at workplace locations and residence addresses can help to reduce traffic emissions by optimization of transport scenarios.

Chapter 7 - The Air Pollution Control Act (APCA) of Taiwan signed in 1975 prescribes the maximum permissible limits of motor vehicle exhausts as well as the monitoring of air pollution, etc. In this study, the concentrations of air pollutants including SO_2, CO, O_3, PM_{10}, NO_2, and non-methane hydrocarbons (NMHC) from background air pollution monitoring stations (BAQMSs) and traffic area air pollution monitoring stations (TAQMSs) were evaluated to comprehend the variations of traffic-related air pollution in Taiwan from 1994 to

2006. The results indicated that the background concentrations of SO_2 increased from 1994, peaked at 1997 and decreased to 4.31 ppb at 2006. The concentrations of traffic area SO_2 decreased from 16.09 ppb to 7.19 ppb during this period. The background concentrations of CO varied between 0.35 ppm and 0.51 ppm, while the concentrations of traffic area CO decreased from 5.20 ppm to 1.17 ppm from 1994 to 2006. The background concentrations of O_3 increased from 1994, peaked at 2004 (34.10 ppb) and maintained at 33.51 ppb at 2006. The concentrations of traffic area O_3 varied between 18.70 ppb and 25.44 ppb during this period. From 1994 to 2006, the background concentrations of PM_{10} varied between 41.55 $\mu g/m^3$ and 60.73 $\mu g/m^3$, while the concentrations of traffic area PM_{10} decreased from 119.39 $\mu g/m^3$ to 69.62 $\mu g/m^3$. The background concentrations of NO_2 varied between 13.93 ppb and 16.54 ppb, while the concentrations of traffic area NO_2 decreased from 55.85 ppb to 31.68 ppb from 1994 to 2006. The background concentrations of NMHC decreased from 0.61 ppm to 0.11 ppm. The concentrations of traffic area NMHC decreased from 2.52 ppb to 0.72 ppb from 1994 to 2006. From 1994 to 2006, SO_2, CO, PM_{10}, NO_2 and NMHC from TAQMSs decreased by 55.3 %, 77.5 %, 41.7 %, 43.3 % and 71.4 %, respectively. Contrarily, the concentrations of O_3 increased by 17.2 %. The traffic area air quality was improved as the transportation loading increased from 1994. Actually, many efforts including vehicle exhaust emission regulation, fuel and mechanical improvement for cars have been made after the signature of APCA in 1975. But according to the analysis, transportation was the major source of SO_2, CO, PM_{10}, NO_2 and NMHC. For further improving the air quality in Taiwan, TEPA shall adopt stricter regulations for traffic area.

In: Air Pollution Emissions
Editors: Daniel G. Vasiliev and Robert A. Kask

ISBN: 978-1-62100-453-0

Chapter 1

OPPORTUNITIES FOR SUSTAINABLE ENERGY DEVELOPMENT AND APPLICATIONS IN BUILDINGS ENVIRONMENT AND AIR POLLUTION CONTROLS

Abdeen Mustafa Omer
Energy Research Institute (ERI), Khartoum, Sudan

ABSTRACT

People to rely upon oil for primary energy and this for a few more decades. Other conventional sources may be more enduring, but are not without serious disadvantages. The renewable energy resources are particularly suited for the provision of rural power supplies and a major advantage is that equipment such as flat plate solar driers, wind machines, etc., can be constructed using local resources and without the advantage results from the feasibility of local maintenance and the general encouragement such local manufacture gives to the build up of small-scale rural based industry. This chapter comprises a comprehensive review of energy sources, the environment and sustainable development. It includes the renewable energy technologies, energy efficiency systems, energy conservation scenarios, energy savings in greenhouses environment and other mitigation measures necessary to reduce climate change. This theme gives some examples of small-scale energy converters, nevertheless it should be noted that small conventional, i.e., engines are currently the major source of power in rural areas and will continue to be so for a long time to come. There is a need for some further development to suit local conditions, to minimise spares holdings, to maximise interchangeability both of engine parts and of the engine application. Emphasis should be placed on full local manufacture. It is concluded that renewable environmentally friendly energy must be encouraged, promoted, implemented and demonstrated by full-scale plan especially for use in remote rural areas.

Keywords: Renewable energy technologies, energy efficiency, sustainable development, emissions, environment

1. Introduction

Power from natural resources has always had great appeal. Coal is plentiful, though there is concern about despoliation in winning it and pollution in burning it. Nuclear power has been developed with remarkable timeliness, but is not universally welcomed, construction of the plant is energy-intensive and there is concern about the disposal of its long-lived active wastes. Barrels of oil, lumps of coal, even uranium come from nature but the possibilities of almost limitless power from the atmosphere and the oceans seem to have special attraction. The wind machine provided an early way of developing motive power. The massive increases in fuel prices over the last years have however, made any scheme not requiring fuel appear to be more attractive and to be worth reinvestigation. In considering the atmosphere and the oceans as energy sources the four main contenders are wind power, wave power, tidal and power from ocean thermal gradients. The sources to alleviate the energy situation in the world are sufficient to supply all foreseeable needs. Conservation of energy and rationing in some form will however have to be practised by most countries, to reduce oil imports and redress balance of payments positions. Meanwhile development and application of nuclear power and some of the traditional solar, wind and water energy alternatives must be set in hand to supplement what remains of the fossil fuels.

The encouragement of greater energy use is an essential component of development. In the short-term, it requires mechanisms to enable the rapid increase in energy/capita, while in the long-term it may require the use of energy efficiency without environmental and safety concerns. Such programmes should as far as possible be based on renewable energy resources.

Large-scale, conventional, power plant such as hydropower, has an important part to play in development although it does not provide a complete solution. There is however an important complementary role for the greater use of small-scale, rural based, and power plants. Such plants can be employed to assist development since they can be made locally. Renewable resources are particularly suitable for providing the energy for such equipment and its use is also compatible with the long-term aims.

In compiling energy consumption data one can categorise usage according to a number of different schemes:

- Traditional sector- industrial, transportation, etc.
- End-use- space heating, process steam, etc.
- Final demand- total energy consumption related to automobiles, to food, etc.
- Energy source- oil, coal, etc.
- Energy form at point of use- electric drive, low temperature heat, etc.

2. Renewable Energy Potential

The increased availability of reliable and efficient energy services stimulates new development alternatives (Omer, 2009a). This communication discusses the potential for such integrated systems in the stationary and portable power market in response to the critical need for a cleaner energy technology. Anticipated patterns of future energy use and consequent environmental impacts (acid precipitation, ozone depletion and the greenhouse effect or global warming) are comprehensively discussed in this approach. Throughout the theme several issues relating to renewable energies, environment and sustainable development are examined from both current and future perspectives. It is concluded that renewable environmentally friendly energy must be encouraged, promoted, implemented and demonstrated by full-scale plan especially for use in remote rural areas. Globally, buildings are responsible for approximately 40% of the total world annual energy consumption. Most of this energy is for the provision of lighting, heating, cooling, and air conditioning. Increasing awareness of the environmental impact of CO_2 and NO_x, and CFCs emissions triggered a renewed interest in environmentally friendly cooling, and heating technologies. Under the 1997 Montreal Protocol, governments agreed to phase out chemicals used as refrigerants that have the potential to destroy stratospheric ozone. It was therefore considered desirable to reduce energy consumption and decrease the rate of depletion of world energy reserves and pollution of the environment. One way of reducing building energy consumption is to design buildings, which are more economical in their use of energy for heating, lighting, cooling, ventilation and hot water supply. Passive measures, particularly natural or hybrid ventilation rather than air-conditioning, can dramatically reduce primary energy consumption. However, exploitation of renewable energy in buildings and agricultural greenhouses can, also, significantly contribute towards reducing dependency on fossil fuels. Therefore, promoting innovative renewable applications and reinforcing the renewable energy market will contribute to preservation of the ecosystem by reducing emissions at local and global levels. This will also contribute to the amelioration of environmental conditions by replacing conventional fuels with renewable energies that produce no air pollution or greenhouse gases.

There is strong scientific evidence that the average temperature of the earth's surface is rising. This is a result of the increased concentration of carbon dioxide and other GHGs in the atmosphere as released by burning fossil fuels. This global warming will eventually lead to substantial changes in the world's climate, which will, in turn, have a major impact on human life and the built environment. Therefore, effort has to be made to reduce fossil energy use and to promote green energies, particularly in the building sector. Energy use reductions can be achieved by minimising the energy demand, by rational energy use, by recovering heat and the use of more green energies. This study was a step towards achieving that goal. The adoption of green or sustainable approaches to the way in which society is run is seen as an important strategy in finding a solution to the energy problem. The key factors to reducing and controlling CO_2, which is the major contributor to global warming, are the use of alternative approaches to energy generation and the exploration of how these alternatives are used today and may be used in the future as green energy sources (Omer, 2009b). Even with modest assumptions about the availability of land, comprehensive fuel-wood farming programmes offer significant energy, economic and environmental benefits. These benefits would be dispersed in rural areas where they are greatly needed and can serve as linkages for further rural economic development. The nations as a whole would benefit from savings in foreign exchange, improved energy security, and socio-economic improvements. With a nine-fold increase in forest – plantation cover, a nation's resource base would be greatly improved. The international community would benefit from pollution reduction, climate mitigation, and the increased trading opportunities that arise from new income sources. The non-technical issues, which have recently gained attention, include: (1) Environmental and ecological factors, e.g., carbon sequestration, reforestation and revegetation. (2) Renewables as a CO_2

neutral replacement for fossil fuels. (3) Greater recognition of the importance of renewable energy, particularly modern biomass energy carriers, at the policy and planning levels. (4) Greater recognition of the difficulties of gathering good and reliable renewable energy data, and efforts to improve it. (5) Studies on the detrimental health efforts of biomass energy particularly from traditional energy users. The renewable energy resources are particularly suited for the provision of rural power supplies and a major advantage is that equipment such as flat plate solar driers, wind machines, etc., can be constructed using local resources and without the advantage results from the feasibility of local maintenance and the general encouragement such local manufacture gives to the build up of small-scale rural based industry. This study gives some examples of small-scale energy converters, nevertheless it should be noted that small conventional, i.e., engines are currently the major source of power in rural areas and will continue to be so for a long time to come. There is a need for some further development to suit local conditions, to minimise spares holdings, to maximise interchangeability both of engine parts and of the engine application. Emphasis should be placed on full local manufacture.

The renewable energy resources are particularly suited for the provision of rural power supplies and a major advantage is that equipment such as flat plate solar driers, wind machines, etc., can be constructed using local resources and without the high capital cost of more conventional equipment. Further advantage results from the feasibility of local maintenance and the general encouragement such local manufacture gives to the build up of small scale rural based industry. Table 1 lists the energy sources available. Scientifically, it is difficult to predict the relationship between global temperature and greenhouse gas (GHG) emissions concentrations. The climate system contains many processes that will change if warming occurs. The equipment and infrastructure for energy supply and use are designed with long lifetimes, and the premature turnover of capital stock involves significant costs.

Table 1. Sources of energy

Energy source	Energy carrier	Energy end-use
Vegetation	Fuel-wood	Cooking Water heating Building materials Animal fodder preparation
Oil	Kerosene	Lighting Ignition fires
Dry cells	Dry cell batteries	Lighting Small appliances
Muscle power	Animal power	Transport Land preparation for farming Food preparation (threshing)
Muscle power	Human power	Transport Land preparation for farming Food preparation (threshing)

Currently the 'non-commercial' fuels wood, crop residues and animal dung are used in large amounts in the rural areas of developing countries, principally for heating and cooking; the method of use is highly inefficient. Table 2 presented some renewable applications.

Table 2. Renewable applications

Systems	Applications
Water supply	Rain collection, purification, storage and recycling
Wastes disposal	Anaerobic digestion (CH_4)
Cooking	Methane
Food	Cultivate the 1 hectare plot and greenhouse for four people
Electrical demands	Wind generator
Space heating	Solar collectors
Water heating	Solar collectors and excess wind energy
Control system	Ultimately hardware
Building fabric	Integration of subsystems to cut costs

Table 3 lists the most important of energy needs.
Considerations when selecting power plant include the following:

- Power level- whether continuous or discontinuous.
- Cost- initial cost, total running cost including fuel, maintenance and capital amortised over life.
- Complexity of operation.
- Maintenance and availability of spares.
- Life.
- Suitability for local manufacture.

Table 4 listed methods of energy conversion.

Table 3. Energy needs in rural areas

Transport, e.g., small vehicles and boats
Agricultural machinery, e.g., two-wheeled tractors
Crop processing, e.g., milling
Water pumping
Small industries, e.g., workshop equipment
Electricity generation, e.g., hospitals and schools
Domestic, e.g., cooking, heating, and lighting
Water supply, e.g., rain collection, purification, storage and recycling
Building fabric, e.g., integration of subsystems to cut costs
Wastes disposal, e.g., anaerobic digestion (CH_4)

Table 4. Methods of energy conversion

Muscle power	Man, animals
Internal combustion engines	
Reciprocating	Petrol- spark ignition Diesel- compression ignition Humphrey water piston
Rotating	Gas turbines
Heat engines	
Vapour (Rankine)	
Reciprocating	Steam engine
Rotating	Steam turbine
Gas Stirling (Reciprocating)	Steam engine
Gas Brayton (Rotating)	Steam turbine
Electron gas	Thermionic, thermoelectric
Electromagnetic radiation	Photo devices
Hydraulic engines	Wheels, screws, buckets, and turbines
Wind engines (wind machines)	Vertical axis, and horizontal axis
Electrical/mechanical	Dynamo/alternator, and motor

The household wastes, i.e., for family of four persons, could provide 280 kWh/yr of methane, but with the addition of vegetable wastes from 0.2 ha or wastes from 1 ha growing a complete diet, about 1500 kWh/yr may be obtained by anaerobic digestion (Omer, 2009c). The sludge from the digester may be returned to the land. In hotter climates, this could be used to set up a more productive cycle (Figure 1).

There is a need for greater attention to be devoted to this field in the development of new designs, the dissemination of information and the encouragement of its use. International and government bodies and independent organisations all have a role to play in renewable energy technologies.

Society and industry in Europe and elsewhere are increasingly dependent on the availability of electricity supply and on the efficient operation of electricity systems. In the European Union (EU), the average rate of growth of electricity demand has been about 1.8% per year since 1990 and is projected to be at least 1.5% yearly up to 2030 (Omer, 2009c). Currently, distribution networks generally differ greatly from transmission networks, mainly in terms of role, structure (radial against meshed) and consequent planning and operation philosophies.

However, renewable energy technologies (RETs) have the benefit of being environmentally benign when developed in a sensitive and appropriate way with the full involvement of local communities. In addition, they are diverse, secure, locally based and abundant. In spite of the enormous potential and the multiple benefits, the contribution from renewable energy still lags behind the ambitious claims for it due to the initially high development costs, concerns about local impacts, lack of research funding and poor institutional and economic arrangements. Hence, an approach is needed to integrate renewable energies in a way that meets high building performance requirements.

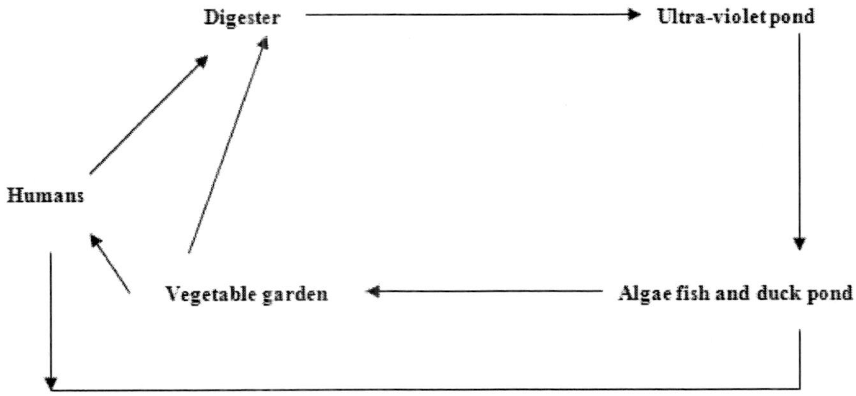

Figure 1. Biomass energy utilisation cycle.

3. Energy Consumption

Over the last decades, natural energy resources such as petroleum and coal have been consumed at high rates. The heavy reliance of the modern economy on these fuels are bound to end, due to their environmental impact, and the fact that conventional sources might eventually run out. The increasing price of oil and instabilities in the oil market led to search for energy substitutes.

In addition to the drain on resources, such an increase in consumption consequences, together with the increased hazards of pollution and the safety problems associated with a large nuclear fission programmes. This is a disturbing prospect. It would be equally unacceptable to suggest that the difference in energy between the developed and developing countries and prudent for the developed countries to move towards a way of life which, whilst maintaining or even increasing quality of life, reduce significantly the energy consumption per capita. Such savings can be achieved in a number of ways:

- Improved efficiency of energy use, for example better thermal insulation, energy recovery, and total energy.
- Conservation of energy resources by design for long life and recycling rather than the short life throwaway product.
- Systematic replanning of our way of life, for example in the field of transport.

Energy ratio is defined as the ratio of:

Energy content of the food product/Energy input to produce the food (1)

A review of the potential range of recyclables is presented in Table 5.

Currently the non-commercial fuelwood, crop residues and animal dung are used in large amounts in the rural areas of developing countries, principally for heating and cooking, the method of use is highly inefficient. As in the developed countries, the fossil fuels are currently of great importance in the developing countries. Geothermal and tidal energy are less important though, of course, will have local significance where conditions are suitable. Nuclear energy sources are included for completeness, but are not likely to make any effective contribution in the rural areas.

Table 5. Summary of material recycling practices in the construction sector

Construction and demolition material	Recycling technology options	Recycling product
Asphalt	Cold recycling: heat generation; Minnesota process; parallel drum process; elongated drum; microwave asphalt recycling system; finfalt; surface regeneration	Recycling asphalt; asphalt aggregate
Brick	Burn to ash, crush into aggregate	Slime burn ash; filling material; hardcore
Concrete	Crush into aggregate	Recycling aggregate; cement replacement; protection of levee; backfilling; filter
Ferrous metal	Melt; reuse directly	Recycled steel scrap
Glass	Reuse directly; grind to powder; polishing; crush into aggregate; burn to ash	Recycled window unit; glass fibre; filling material; tile; paving block; asphalt; recycled aggregate; cement replacement; manmade soil
Masonry	Crush into aggregate; heat to 900°C to ash	Thermal insulating concrete; traditional clay
Non-ferrous metal	Melt	Recycled metal
Paper and cardboard	Purification	Recycled paper
Plastic	Convert to powder by cryogenic milling; clopping; crush into aggregate; burn to ash	Panel; recycled plastic; plastic lumber; recycled aggregate; landfill drainage; asphalt; manmade soil
Timber	Reuse directly; cut into aggregate; blast furnace deoxidisation; gasification or pyrolysis; chipping; moulding by pressurising timber chip under steam and water	Whole timber; furniture and kitchen utensils; lightweight recycled aggregate; source of energy; chemical production; wood-based panel; plastic lumber; geofibre; insulation board

Robinson, 2007.

4. BIOGAS PRODUCTION

Biogas is a generic term for gases generated from the decomposition of organic material. As the material breaks down, methane (CH_4) is produced as shown in Figure 2. Sources that generate biogas are numerous and varied. These include landfill sites, wastewater treatment plants and anaerobic digesters (Omer, 2009d). Landfills and wastewater treatment plants emit biogas from decaying waste. To date, the waste industry has focused on controlling these emissions to our environment and in some cases, tapping this potential source of fuel to power gas turbines, thus generating electricity (Omer, 2009d). The primary components of landfill gas are methane (CH_4), carbon dioxide (CO_2), and nitrogen (N_2). The average concentration of methane is ~45%, CO_2 is ~36% and nitrogen is ~18% (Omer, and Yemen, 2001). Other components in the gas are oxygen (O_2), water vapour and trace amounts of a wide range of non-methane organic compounds (NMOCs). Landfill gas-to-cogeneration projects present a win-win-win situation. Emissions of particularly damaging pollutant are avoided, electricity is generated from a free fuel and heat is available for use locally. At present, most of the energy used to heat buildings, including electrical energy, comes from fossil fuels such as oil and coal. This energy originally came from the sun and was used in the growth of plants such as trees. Then, because of changes in the earth's geology, those ancient forests eventually became a coal seam, an oil field or a natural gas field.

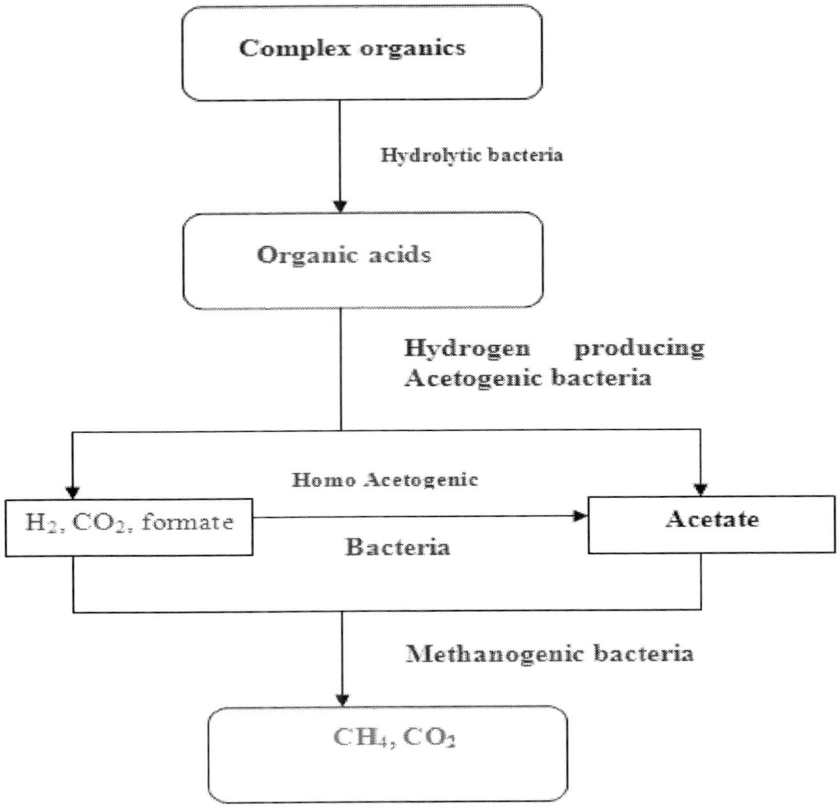

Figure 2. Biogas production process.

5. Wave Power Conversion Devices

The patent literature is full of devices for extracting energy from waves, i.e., floats, ramps, and flaps, covering channels (Swift-Hook, et al, 1975). Small generators driven from air trapped by the rising and falling water in the chamber of a buoy are in use around the world (Swift-Hook, et al, 1975). Wave power is one possibility that has been selected. Figure 3 shows the many other aspects that will need to be covered. A wave power programme would make a significant contribution to energy resources within a relatively short time and with existing technology.

Wave energy has also been in the news recently. There is about 140 megawatts per mile available round British coasts. It could make a useful contribution people needs in the UK. Although very large amounts of power are available in the waves, it is important to consider how much power can be extracted. A few years ago only a few percent efficiency had been achieved. Recently, however, several devices have been studied which have very high efficiencies. Some form of storage will be essential on a second-to-second and minute-to-minute basis to smooth the fluctuations of individual waves and wave's packets but storage from one day to the next will certainly not be economical. This is why provision must be made for adequate standby capacity.

The increased availability of reliable and efficient energy services stimulates new development alternatives. This study discusses the potential for such integrated systems in the stationary and portable power market in response to the critical need for a cleaner energy technology. Anticipated patterns of future energy use and consequent environmental impacts (acid precipitation, ozone depletion and the greenhouse effect or global warming) are comprehensively discussed in this theme. Throughout the theme several issues relating to renewable energies, environment and sustainable development are examined from both current and future perspectives. It is concluded that renewable environmentally friendly energy must be encouraged, promoted, implemented and demonstrated by full-scale plan especially for use in remote rural areas. Globally, buildings are responsible for approximately 40% of the total world annual energy consumption. Most of this energy is for the provision of lighting, heating, cooling, and air conditioning. Increasing awareness of the environmental impact of CO_2 and NO_x, and CFCs emissions triggered a renewed interest in environmentally friendly cooling, and heating technologies. Under the 1997 Montreal Protocol, governments agreed to phase out chemicals used as refrigerants that have the potential to destroy stratospheric ozone. It was therefore considered desirable to reduce energy consumption and decrease the rate of depletion of world energy reserves and pollution of the environment. One way of reducing building energy consumption is to design buildings, which are more economical in their use of energy for heating, lighting, cooling, ventilation and hot water supply. Passive measures, particularly natural or hybrid ventilation rather than air-conditioning, can dramatically reduce primary energy consumption. However, exploitation of renewable energy in buildings and agricultural greenhouses can, also, significantly contribute towards reducing dependency on fossil fuels. Therefore, promoting innovative renewable applications and reinforcing the renewable energy market will contribute to preservation of the ecosystem by reducing emissions at local and global levels. This will also contribute to the amelioration of environmental conditions by replacing conventional fuels with renewable energies that produce no air pollution or greenhouse gases. The provision of good indoor environmental quality while achieving energy and cost efficient operation of the heating, ventilating and air-conditioning (HVAC) plants in buildings represents a multi variant problem. The comfort of building occupants is dependent on many environmental parameters including air speed, temperature, relative humidity and quality in addition to lighting and noise. The overall objective is to provide a high level of building performance (BP), which can be defined as indoor environmental quality (IEQ), energy efficiency (EE) and cost efficiency (CE).

- Indoor environmental quality is the perceived condition of comfort that building occupants experience due to the physical and psychological conditions to which they are exposed by their surroundings. The main physical parameters affecting IEQ are air speed, temperature, relative humidity and quality.
- Energy efficiency is related to the provision of the desired environmental conditions while consuming the minimal quantity of energy.
- Cost efficiency is the financial expenditure on energy relative to the level of environmental comfort and productivity that the building occupants attained. The overall cost efficiency can be improved by improving the indoor environmental quality and the energy efficiency of a building.

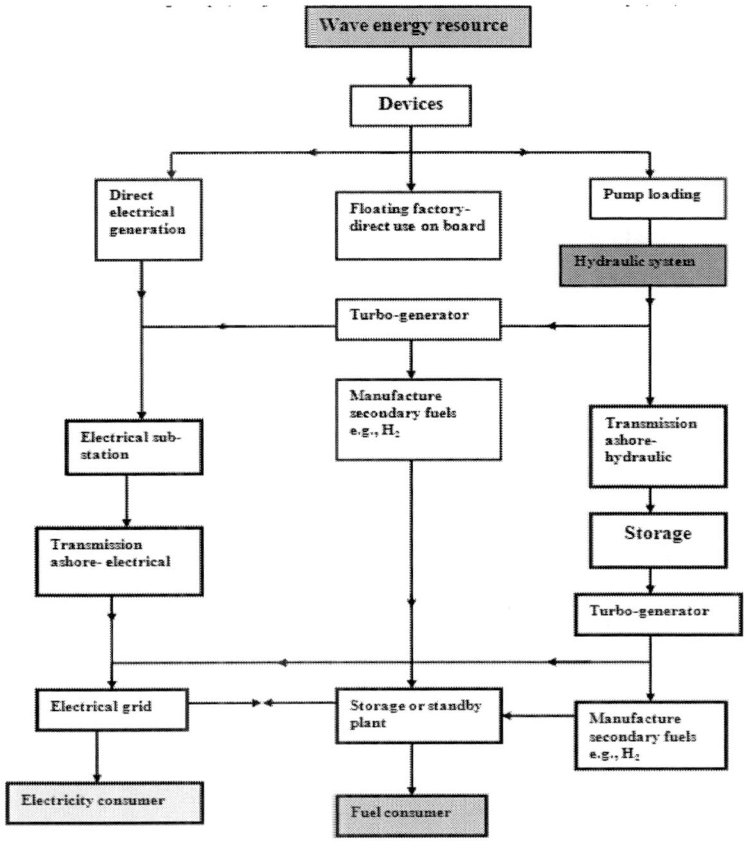

Figure 3. Possible systems for exploiting wave power, each element represents an essential link in the chain from sea waves to consumer.

An approach is needed to integrate renewable energies in a way to meet high building performance. However, because renewable energy sources are stochastic and geographically diffuse, their ability to match demand is determined by adoption of one of the following two approaches: the utilisation of a capture area greater than that occupied by the community to be supplied, or the reduction of the community's energy demands to a level commensurate with the locally available renewable resources.

6. ETHANOL PRODUCTION

Alternative fuels were defined as methanol, ethanol, natural gas, propane, hydrogen, coal-derived liquids, biological material and electricity production (Sims, 2007). The fuel pathways currently under development for alcohol fuels are shown in Figure 4. The production of agricultural biomass and its exploitation for energy purposes can contribute to alleviate several problems, such as the dependence on import of energy products, the production of food surpluses, the pollution provoked by the use of fossil fuels, the abandonment of land by farmers and the connected urbanisation. Biomass is not at the moment competitive with mineral oil, but, taking into account also indirect costs and giving a value to the aforementioned advantages, public authorities at national and international level can spur its production and use by incentives of different nature. In order to address the problem of inefficiency, research centres around the world have investigated the viability of converting the resource to a more useful form, namely solid briquettes and fuel gas (Sims, 2007) (Figure 5).

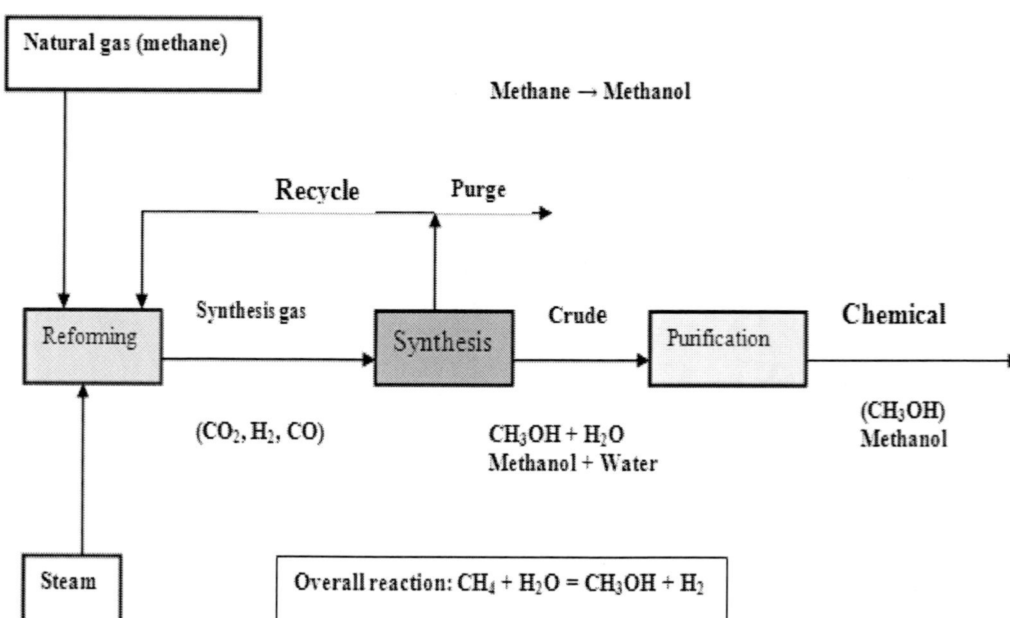

Figure 4. Schematic process flowsheet.

The main advantages are related to energy, agriculture and environment problems, are foreseeable both at regional level and at worldwide level and can be summarised as follows:

- Reduction of dependence on import of energy and related products.
- Reduction of environmental impact of energy production (greenhouse effect, air pollution, and waste degradation).
- Substitution of food crops and reduction of food surpluses and of related economic burdens, and utilisation of marginal lands and of set aside lands.
- Reduction of related socio-economic and environmental problems (soil erosion, urbanisation, landscape deterioration, etc.).
- Development of new know-how and production of technological innovation.

For a northern European climate, which is characterised by an average annual solar irradiance of 150 Wm^{-2}, the mean power production from a photovoltaic component of 13% conversion efficiency is approximately 20 Wm^{-2}. For an average wind speed of 5 ms^{-1}, the power produced by a micro wind turbine will be of a similar order of magnitude, though with a different profile shape. In the UK, for example, a typical office building will have a demand in the order of 300 $kWhm^{-2}yr^{-1}$. This translates into approximately 50 Wm^{-2} of façade, which is twice as much as the available renewable energies. Thus, the aim is to utilise energy efficiency measures in order to reduce the overall energy consumption and adjust the demand profiles to be met by renewable energies. For instance, this approach can be applied to greenhouses, which use solar energy to provide indoor environmental quality. The greenhouse effect is one result of the differing properties of heat radiation when it is generated at different temperatures. Objects inside the greenhouse, or any other building, such as plants, re-radiate the heat or absorb it. Because the objects inside the greenhouse are at a lower temperature than the sun, the re-radiated heat is of longer wavelengths, and cannot penetrate the glass. This re-radiated heat is trapped and causes the temperature inside the greenhouse to rise. Note that the atmosphere surrounding the earth, also, behaves as a large greenhouse around the world. Changes to the gases in the atmosphere, such as increased carbon dioxide content from the burning of fossil fuels, can act like a layer of glass and reduce the quantity of heat that the planet earth would otherwise radiate back into space. This particular greenhouse effect, therefore, contributes to global warming. The application of greenhouses for plants growth can be considered one of the measures in the success of solving this problem. Maximising the efficiency gained from a greenhouse can be achieved using various approaches, employing different techniques that could be applied at the design, construction and operational stages.

Biomass resources play a significant role in energy supply in all developing countries. Biomass resources should be divided into residues or dedicated resources, the latter including firewood and charcoal can also be produced from forest residues. Ozone (O_3) is a naturally occurring molecule that consists of three oxygen atoms held together by the bonding of the oxygen atoms to each other. The effects of the chlorofluorocarbons (CFCs) molecule can last for over a century. This reaction is shown in Figure 6.

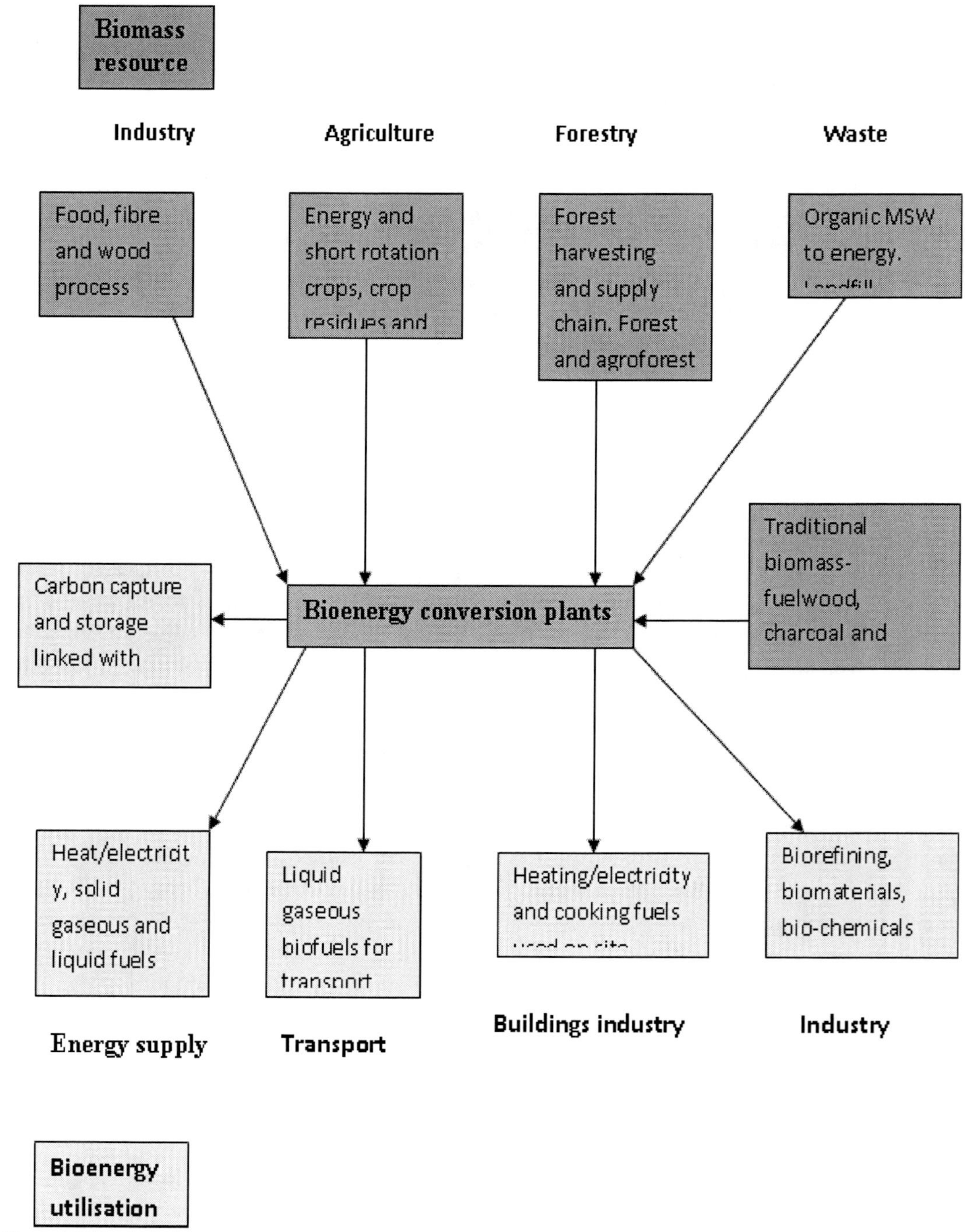

Sims, 2007.

Figure 5. Biomass resources from several sources is converted into a range of products for use by transport, industry and building sectors.

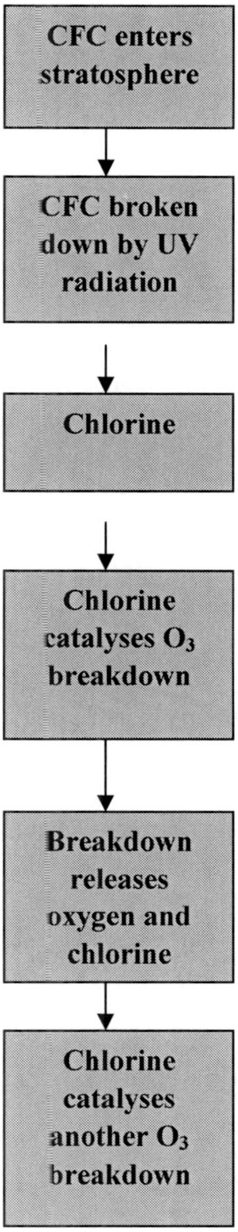

Trevor, 2007.

Figure 6. The process of ozone depletion.

It is a common misconception that the reason for recycling old fridge is to recover the liquid from the cooling circuit at the back of the unit. The insulating foams used inside some fridges act as sinks of CFCs- the gases having been used as blowing agents to expand the foam during fridge manufacture. Although the use of ozone depleting chemicals in the foam in fridges has declined in the West, recyclers must consider which strategy to adopt to deal with the disposal problem they still present each year. It is common practice to dispose of this waste wood in landfill where it slowly degraded and takes up valuable void space. This wood is a good source of energy and is an alternative to energy crops. Agricultural wastes are abundantly available globally and can be converted to energy and useful chemicals by a number of microorganisms. The success of promoting any technology depends on careful planning, management, implementation, training and monitoring. Main features of gasification project are:

- Networking and institutional development/strengthening.
- Promotion and extension.
- Construction of demonstration projects.
- Research and development; and training and monitoring.

7. Biomass CHP

Combined heat and power (CHP) installations are quite common in greenhouses, which grow high-energy, input crops (e.g., salad vegetables, pot plants, etc.). Scientific assumptions for a short-term energy strategy suggest that the most economically efficient way to replace the thermal plants is to modernise existing power plants to increase their energy efficiency and to improve their environmental performance. However, utilisation of wind power and the conversion of gas-fired CHP plants to biomass would significantly reduce the dependence on imported fossil fuels. Although a lack of generating capacity is forecasted in the long-term, utilisation of the existing renewable energy potential and the huge possibilities for increasing energy efficiency are sufficient to meet future energy demands in the short-term.

A total shift towards a sustainable energy system is a complex and long process, but is one that can be achieved within a period of about 20 years. Implementation will require initial investment, long-term national strategies and action plans. However, the changes will have a number of benefits including: a more stable energy supply than at present, and major improvement in the environmental performance of the energy sector, and certain social benefits. A national vision (Omer, 2009d) used a methodology and calculations based on computer modelling that utilised:

- Data from existing governmental programmes.
- Potential renewable energy sources and energy efficiency improvements and assumptions for future economy growth.
- Information from studies and surveys on the recent situation in the energy sector.

In addition to realising the economic potential identified by the National Energy Savings Programme, a long-term effort leading to a 3% reduction in specific electricity demand per year after 2020 is proposed. This will require further improvements in building codes, and continued information on energy efficiency. The environmental Non Governmental Organisations (NGOs) are urging the government to adopt sustainable development of the energy sector by:

- Diversifying of primary energy sources to increase the contribution of renewable and local energy resources in the total energy balance.
- Implementing measures for energy efficiency increase at the demand side and in the energy transformation sector.

The price of natural gas is set by a number of market and regulatory factors that include: supply and demand balance and market fundamentals, weather, pipeline availability and deliverability, storage inventory, new supply sources, prices of other energy alternatives and regulatory issues and uncertainty. Classic management approaches to risk are well documented and used in many industries. This includes the following four broad approaches to risk:

- Avoidance includes not performing an activity that could carry risk. Avoidance may seem the answer to all risks, but avoiding risks also means losing out on potential gain.
- Mitigation/reduction involves methods that reduce the severity of potential loss.
- Retention/acceptance involves accepting the loss when it occurs. Risk retention is a viable strategy for small risks. All risks that are not avoided or transferred are retained by default.
- Transfer means causing another party to accept the risk, typically by contract.

Methane is a primary constituent of landfill gas (LFG) and a potent greenhouse gas (GHG) when released into the atmosphere. Globally, landfills are the third largest anthropogenic emission source, accounting for about 13% of methane emissions or over 818 million tones of carbon dioxide equivalent ($MMTCO_2e$) (Brain, and Mark 2007) as shown in Figures 7-9.

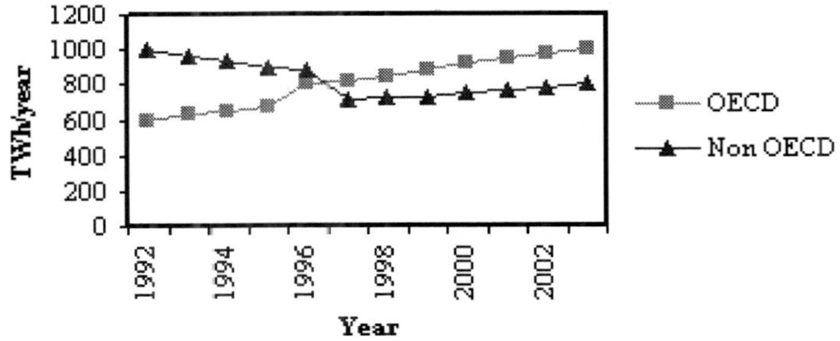

IEA, 2007.

Figure 7. Global CHP trends from 1992-2003.

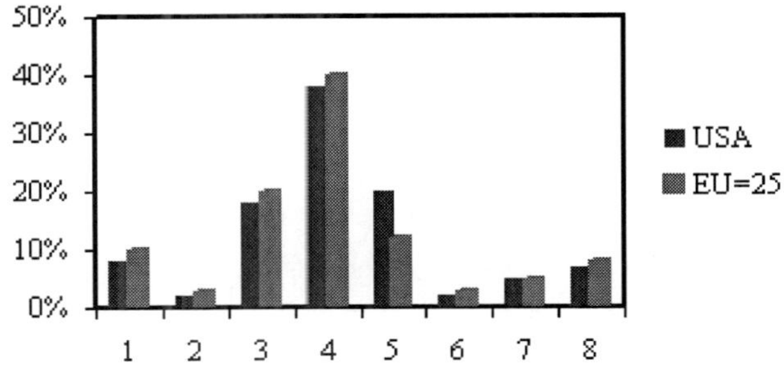

1 Food, 2 Textile, 3 Pulp & paper, 4 Chemicals, 5 Refining, 6 Minerals, 7 Primary metals, and 8 others
IEA, 2007.

Figure 8. Distribution of industrial CHP capacity in the EU and USA.

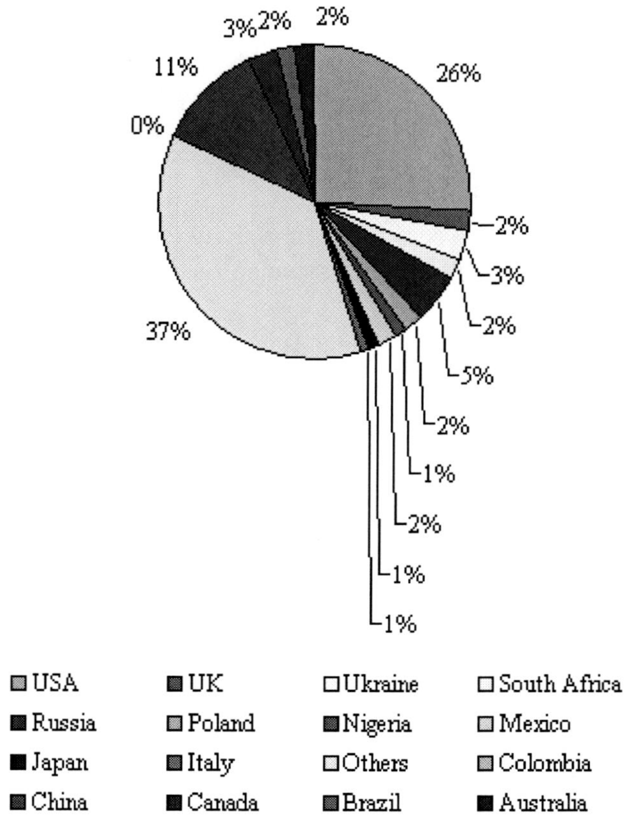

Brain, and Mark 2007.

Figure 9. World landfill methane emissions (MMTCO2e).

8. Geothermal Energy

Geothermal steam has been used in volcanic regions in many countries to generate electricity. The use of geothermal energy involves the extraction of heat from rocks in the outer part of the earth. It is relatively unusual for the rocks to be sufficiently hot at shallow depth for this to be economically attractive. Virtually all the areas of present geothermal interest are concentrated along the margins of the major tectonic plates, which form the surface of the earth. The forced or natural circulation of water through permeable hot rock conventionally extracts heat.

There are various practical difficulties and disadvantages associated with the use of geothermal power:

Transmission: geothermal power has to be used where it is found. In Iceland it has proved feasible to pipe hot water 20 km in insulated pipes but much shorter distances are preferred.

Environmental problems: these are somewhat variable and are usually not great. Perhaps the most serious is the disposal of warm high salinity water where it cannot be reinjected or purified. Dry steam plants tend to be very noisy and there is releases of small amounts of methane, hydrogen, nitrogen, amonia and hydrogen sulphide and of these the latter presents the main problem.

The geothermal fluid is often highly chemically corrosive or physically abrassive as the result of the entrained solid matter it carries. This may entail special plant design problems and unusually short operational lives for both the holes and the installations they serve.

Because the useful rate of heat extraction from a geothermal field is in nearly all cases much higher than the rate of conduction into the field from the underlying rocks, the mean temperatures of the field is likely to fall during exploitation. In some low rainfall areas there may also be a problem of fluid depletion. Ideally, as much as possible of the geothermal fluid should be reinjected into the field. However, this may involve the heavy capital costs of large condensation installations. Occasionally, the salinity of the fluid available for reinjection may be so high (as a result of concentration by boiling) that is unsuitable for reinjection into ground. Ocasionally, the impurities can be precipitated and used but this has not generally proved commercially attractive.

World capacity of geothermal energy is growing at a rate of 2.5% per year from a 2005 level of 28.3 GW (Rawlings, 1999). The GSHPs account for approximately 54% of this capacity almost all of it in the North America and Europe (Rawlings, 1999). The involvement of the UK is minimal with less than 0.04% of world capacity and yet is committed to substantial reduction in carbon emission beyond the 12.5% Kyoto obligation to be achieved by 2012. The GSHPs offer a significant potential for carbon reduction and it is therefore expected that the market for these systems will rise sharply in the UK in the immediate years ahead given to low capacity base at present.

There are numerous ways of harnessing low-grade heat from the ground for use as a heat pump source or air conditioning sink. For small applications (residences and small commercial buildings) horizontal ground loop heat exchangers buried typically at between 1 m and 1.8 m below the surface can be used provided that a significant availability of land surrounding the building can be exploited which tends to limit these applications to rural settings.

Heat generation within the earth is approximately 2700 GW, roughly an order of magnitude greater than the energy associated with the tides but about four orders less than that received by the earth from the sun (Oxburgh, 1975).

Temperature distributions within the earth depend on:

- The abundance and distribution of heat producing elements within the earth.
- The mean surface temperature (which is controlled by the ocean/atmosphere system).
- The thermal properties of the earth's interior and their lateral and radial variation.
- Any movements of fluid or solid rock materials occurring at rates of more than a few millimetres per year.

Of these four factors the first two are of less importance from the point of view of geothermal energy. Mean surface temperatures range between 0-30°C and this variation has a small effect on the useable enthalpy of any flows of hot water. Although radiogenic heat production in rocks may vary by three orders of magnitude, there is much less variation from place to place in the integrated heat production with depth. The latter factors, however, are of great importance and show a wide range of variation. Their importance is clear from the relationship:

$$\beta = q/k \quad (2)$$

where:

β is the thermal gradient for a steady state (°C/km), q is the heat flux (10^{-6} cal cm^{-2} sec^{-1}) and k is the thermal conductivity (cal cm^{-1} sec^{-1} $°C^{-1}$).

The first requirement of any potential geothermal source region is that β being large, i.e., that high rock temperatures occur at shallow depth. Beta will be large if either q is large or k is small or both. By comparison with most everyday materials, rocks are poor conductors of heat and values of conductivity may vary from 2 x 10^{-3} to 10^{-2} cal cm^{-1} sec^{-1} $°C^{-1}$. The mean surface heat flux from the earth is about 1.5 heat flow units (1 HFU = 10^{-6} cal cm^{-2} sec^{-1}) (Oxburgh, 1975). Rocks are also very slow respond to any temperature change to which they are exposed, i.e., they have a low thermal diffusivity:

$$K = k/\rho C_p \quad (3)$$

where:

K is thermal diffusivity; ρ and C_p are density and specific heat respectively.

These values are simple intended to give a general idea of the normal range of geothermal parameters (Table 6). In volcanic regions, in particular, both q and β can vary considerably and the upper values given are somewhat nominal.

Table 6. Values of geothermal parameters

Parameter	Lower	Average	Upper
q (HFU)	0.8	1.5	3.0 (non volcanic) ≈100 (volcanic)
k =cal cm^{-2} sec-1 $°C^{-1}$	$2x10^{-3}$	$6x10^{-3}$	$12x10^{-3}$
β =°C/km	8	20	60 (non volcanic) ≈300 (volcanic)

9. Landfill Gas

Landfill gas (LFG) is currently extracted at over 1200 landfills worldwide for a variety of energy purposes (Table 7), such as:

- Creating pipeline quality gas or an alternative fuel for vehicles.
- Processing the LFG to make it available as an alternative fuel to local industrial or commercial customers.

- Generation of electricity with engines, turbines, micro-turbines and other emerging technologies.

In terms of solid waste management policy, many NGOs have changed drastically in the past ten years from a mass production and mass consumption society to 'material-cycle society' (Abdeen, 2008). In addition to national legislation, municipalities are legally obliged to develop a plan for handling the municipal solid waste (MSW) generated in administrative areas. Such plans contain:

- Estimates of future waste volume.
- Measures to reduce waste.
- Measures to encourage source separation.
- A framework for solid waste disposal and the construction and management of solid waste management facilities.

Landfilling is in the least referred tier of the hierarchy of waste management options: waste minimisation, reuse and recycling, incineration with energy recovery, and optimised final disposal. The key elements are as follows: construction impacts, atmospheric emissions, noise, water quality, landscape, visual impacts, socio economics, ecological impacts, traffic, solid waste disposal and cultural heritage. The move towards a de-carbonised world, driven partly by climate science and partly by the business opportunities it offers, will need the promotion of environmentally friendly alternatives, if an acceptable stabilisation level of atmospheric carbon dioxide is to be achieved.

Table 7. Types of LFG implemented recently worldwide

Landfill caps ❑ Soil caps ❑ Clay caps ❑ Geo-membrane caps LFG destruction ❑ Flares - Candlestick - Enclosed	Electricity generation ❑ Reciprocating engines ❑ Combustion turbines ❑ Micro-turbines ❑ Steam turbines ❑ Fuel cells CHP ❑ Turbines	Fuel production ❑ Medium BTU gas ❑ High BTU gas ❑ Liquefied methane Thermal generation ❑ Boilers ❑ Kilns ❑ Greenhouse heaters ❑ Leachate evaporators

10. Energy Efficiency

Energy efficiency is the most cost-effective way of cutting carbon dioxide emissions and improvements to households and businesses. It can also have many other additional social, economic and health benefits, such as warmer and healthier homes, lower fuel bills and company running costs and, indirectly, jobs. Britain wastes 20 per cent of its fossil fuel and electricity use. This implies that it would be cost-effective to cut £10 billion a year off the collective fuel bill and reduce CO_2 emissions by some 120 million tones. Yet, due to lack of good information and advice on energy saving, along with the capital to finance energy efficiency improvements, this huge potential for reducing energy demand is not being realised. Traditionally, energy utilities have been essentially fuel providers and the industry has pursued profits from increased volume of sales. Institutional and market arrangements have favoured energy consumption rather than conservation. However, energy is at the centre of the sustainable development paradigm as few activities affect the environment as much as the continually increasing use of energy. Most of the used energy depends on finite resources, such as coal, oil, gas and uranium. In addition, more than three quarters of the world's consumption of these fuels is used, often inefficiently, by only one quarter of the world's population. Without even addressing these inequities or the precious, finite nature of these resources, the scale of environmental damage will force the reduction of the usage of these fuels long before they run out.

Throughout the energy generation process there are impacts on the environment on local, national and international levels, from opencast mining and oil exploration to emissions of the potent greenhouse gas carbon dioxide in ever increasing concentration. Recently, the world's leading climate scientists reached an agreement that human activities, such as burning fossil fuels for energy and transport, are causing the world's temperature to rise. The Intergovernmental Panel on Climate Change has concluded that ''the balance of evidence suggests a discernible human influence on global climate''. It predicts a rate of warming greater than any one seen in the last 10,000 years, in other words, throughout human history. The exact impact of climate change is difficult to predict and will vary regionally. It could, however, include sea level rise, disrupted agriculture and food supplies and the possibility of more freak weather events such as hurricanes and droughts. Indeed, people already are waking up to the financial and social, as well as the environmental, risks of unsustainable energy generation methods that represent the costs of the impacts of climate change, acid rain and oil spills. Techniques considered are hybrid (controlled natural and mechanical) ventilation including night ventilation, thermo-active building mass systems with free cooling in a cooling tower, and air intake via ground heat exchangers. Special emphasis is put on ventilation concepts utilising ambient energy from air ground and other renewable energy sources, and on the interaction with heating and cooling. The insurance industry, for example, concerned about the billion dollar costs of hurricanes and floods, has joined sides with environmentalists to lobby for greenhouse gas emissions reduction. Friends of the earth are campaigning for a more sustainable energy policy, guided by the principal of environmental protection and with the objectives of sound natural resource management and long-term energy security. The key priorities of such an energy policy must be to reduce fossil fuel use, move away from nuclear power, improve the efficiency with which energy is used and increase the amount of energy obtainable from sustainable, and renewable energy sources. Efficient energy use has never been more crucial than it is today, particularly with the prospect of the imminent introduction of the climate change levy (CCL). Establishing an energy use action plan is the essential foundation to the elimination of energy waste. A logical starting point is to carry out an energy audit that enables the assessment of the energy use and determine what actions to take. The actions are best categorised by splitting measures into the following three general groups:

1) High Priority/Low Cost

These are normally measures, which require minimal investment and can be implemented quickly. The followings are some examples of such measures:

- Good housekeeping, monitoring energy use and targeting waste-fuel practices and adjusting controls to match requirements.
- Improved greenhouse space utilisation.
- Small capital item time switches, thermostats, etc.
- Carrying out minor maintenance and repairs.
- Staff education and training.
- Ensuring that energy is being purchased through the most suitable tariff or contract arrangements.

2) Medium Priority/Medium Cost

Measures, which, although involve little or no design, involve greater expenditure and can take longer to implement. Examples of such measures are listed below:

- New or replacement controls.
- Greenhouse component alteration, e.g., insulation, sealing glass joints, etc.
- Alternative equipment components, e.g., energy efficient lamps in light fittings, etc.

3) Long Term/High Cost

These measures require detailed study and design. They can be best represented by the followings:

- Replacing or upgrading of plant and equipment.
- Fundamental redesign of systems, e.g., CHP installations.

This process can often be a complex experience and therefore the most cost-effective approach is to employ an energy specialist to help.

11. Policy Recommendations for a Sustainable Energy Future

Sustainability is regarded as a major consideration for both urban and rural development. People have been exploiting the natural resources with no consideration to the effects, both short-term (environmental) and long-term (resources crunch). It is also felt that knowledge and technology have not been used effectively in utilising energy resources. Energy is the vital input for economic and social development of any country. Its sustainability is an important factor to be considered. The urban areas depend, to a large extent, on commercial energy sources. The rural areas use non-commercial sources like firewood and agricultural wastes. With the present day trends for improving the quality of life and sustenance of mankind, and environmental issues are considered highly important. In this context, the term energy loss has no significant technical meaning. Instead, the exergy loss has to be considered, as destruction of exergy is possible. Hence, exergy loss minimisation will help in sustainability.

The development of a renewable energy in a country depends on many factors. Those important to success are listed below:

1) Motivation of the Population

The population should be motivated towards awareness of high environmental issues, and rational use of energy in order to reduce cost. Subsidy programme should be implemented as incentives to install biomass energy plants. In addition, image campaigns to raise awareness of renewable technology.

2) Technical Product Development

To achieve technical development of biomass energy technologies the following should be addressed:

- Increasing the longevity and reliability of renewable technology.
- Adapting renewable technology to household technology (hot water supply).
- Integration of renewable technology in heating technology.
- Integration of renewable technology in architecture, e.g., in the roof or façade.
- Development of new applications, e.g., solar cooling.
- Cost reduction.

3) Distribution and Sales

Commercialisation of biomass energy technology requires:

- Inclusion of renewable technology in the product range of heating trades at all levels of the distribution process (wholesale, retail, etc.).
- Building distribution nets for renewable technology.
- Training of personnel in distribution and sales.
- Training of field sales force.

4) Consumer Consultation and Installation

To encourage all sectors of the population to participate in adoption of biomass energy technologies, the following has to be realised:

- Acceptance by craftspeople, and marketing by them.
- Technical training of craftspeople, initial and follow-up training programmes.
- Sales training for craftspeople.
- Information material to be made available to craftspeople for consumer consultation.

5) Projecting and Planning

Successful application of biomass technologies also require:

- Acceptance by decision makers in the building sector (architects, house technology planners, etc.).
- Integration of renewable technology in training.
- Demonstration projects/architecture competitions.
- Biomass energy project developers should prepare to participate in the carbon market by:
    - Ensuring that renewable energy projects comply with Kyoto Protocol requirements.
    - Quantifying the expected avoided emissions.
    - Registering the project with the required offices.
    - Contractually allocating the right to this revenue stream.

- Other ecological measures employed on the development include:
    - Simplified building details.
    - Reduced number of materials.
    - Materials that can be recycled or reused.
    - Materials easily maintained and repaired.
    - Materials that do not have a bad influence on the indoor climate (i.e., non-toxic).
    - Local cleaning of grey water.
    - Collecting and use of rainwater for outdoor purposes and park elements.
    - Building volumes designed to give maximum access to neighbouring park areas.
    - All apartments have visual access to both backyard and park.

6) Energy Saving Measures

The following energy saving measures should also be considered:

- Building integrated solar PV system.
- Day-lighting.

- Ecological insulation materials.
- Natural/hybrid ventilation.
- Passive cooling, and passive solar heating.
- Solar heating of domestic hot water.
- Utilisation of rainwater for flushing.

Improving access for rural and urban low-income areas in developing countries must be through energy efficiency and renewable energies. Sustainable energy is a prerequisite for development. Energy-based living standards in developing countries, however, are clearly below standards in developed countries. Low levels of access to affordable and environmentally sound energy in both rural and urban low-income areas are therefore a predominant issue in developing countries. In recent years many programmes for development aid or technical assistance have been focusing on improving access to sustainable energy, many of them with impressive results.

Apart from success stories, however, experience also shows that positive appraisals of many projects evaporate after completion and vanishing of the implementation expert team. Altogether, the diffusion of sustainable technologies such as energy efficiency and renewable energies for cooking, heating, lighting, electrical appliances and building insulation in developing countries has been slow.

Energy efficiency and renewable energy programmes could be more sustainable and pilot studies more effective and pulse releasing if the entire policy and implementation process was considered and redesigned from the outset. New financing and implementation processes are needed, which allow reallocating financial resources and thus enabling countries themselves to achieve a sustainable energy infrastructure. The links between the energy policy framework, financing and implementation of renewable energy and energy efficiency projects have to be strengthened and capacity building efforts are required.

12. Environmental Aspects of Energy Conversion and Use

Environment has no precise limits because it is in fact a part of everything. Indeed, environment is, as anyone probably already knows, not only flowers blossoming or birds singing in the spring, or a lake surrounded by beautiful mountains. It is also human settlements, the places where people live, work, rest, the quality of the food they eat, the noise or silence of the street they live in. Environment is not only the fact that our cars consume a good deal of energy and pollute the air, but also, that we often need them to go to work and for holidays. Obviously man uses energy just as plants, bacteria, mushrooms, bees, fish and rats do. Man largely uses solar energy- food, hydropower, wood- and thus participates harmoniously in the natural flow of energy through the environment. But man also uses oil, gas, coal and nuclear power. By using such sources of energy, man is thus modifying his environment.

The atmospheric emissions of fossil fuelled installations are mosty aldehydes, carbon monoxide, nitrogen oxides, sulpher oxides and particles (i.e., ash) as well as carbon dioxide. Table 8 shows estimates include not only the releases occuring at the power plant itself but also cover fuel extraction and treatment, as well as the storage of wastes and the aea of land required for operations. Table 9 shows energy consumption in different regions of the world. Low energy design should not be considered in isolation, and in fact, it is a measure, which should work in harmony with other environmental objectives. Hence, building energy study provides opportunities not only for identifying energy and cost savings, but also for examining the indoor and outdoor environment.

Table 8. Annual greenhouse emissions from different sources of power plants

Primary source of energy	Emissions		Waste (x 10^3 metric tones)	Area (km^2)
	Atmosphere	Water		
Coal	380	7-41	60-3000	120
Oil	70-160	3-6	negligible	70-84
Gas	24	1	-	84
Nuclear	6	21	2600	77

Table 9. Energy consumption in different continents

Region	Population (millions)	Energy (Watt/m^2)
Africa	820	0.54
Asia	3780	2.74
Central America	180	1.44
North America	335	0.34
South America	475	0.52
Western Europe	445	2.24
Eastern Europe	130	2.57
Oceania	35	0.08
Russia	330	0.29

13. Greenhouses Environment

Greenhouse cultivation is one of the most absorbing and rewarding forms of gardening for anyone who enjoys growing plants. The enthusiastic gardener can adapt the greenhouse climate to suit a particular group of plants, or raise flowers, fruit and vegetables out of their natural season. The greenhouse can also be used as an essential garden tool, enabling the keen amateur to expand the scope of plants grown in the garden, as well as save money by raising their own plants and vegetables. There was a decline in large private greenhouses during the two world wars due to a shortage of materials for their construction and fuel to heat them. However, in the 1950s mass-produced, small greenhouses became widely available at affordable prices and were used mainly for raising plants (John, 1993). Also, in recent years, the popularity of conservatories attached to the house has soared. Modern double-glazing panels can provide as much insulation as a brick wall to create a comfortable living space, as well as provide an ideal environment in which to grow and display tender plants.

The comfort in a greenhouse depends on many environmental parameters. These include temperature, relative humidity, air quality and lighting. Although greenhouse and conservatory originally both meant a place to house or conserve greens (variegated hollies, cirrus, myrtles and oleanders), a greenhouse today implies a place in which plants are raised while conservatory usually describes a glazed room where plants may or may not play a significant role. Indeed, a greenhouse can be used for so many different purposes. It is, therefore, difficult to decide how to group the information about the plants that can be grown inside it.

Throughout the world urban areas have increased in size during recent decades. About 50% of the world's population and approximately 76% in the more developed countries are urban dwellers (UN, 2001). Even though there is an evidence to suggest that in many 'advanced' industrialised countries there has been a reversal in the rural-to-urban shift of populations, virtually all population growth expected between 2000 and 2030 will be concentrated in urban areas of the world. With an expected annual growth of 1.8%, the world's urban population will double in 38 years (UN, 2001). This represents a serious contributing to the potential problem of maintaining the required food supply. Inappropriate land use and management, often driven by intensification resulting from high population pressure and market forces, is also a threat to food availability for domestic, livestock and wildlife use. Conversion to cropland and urban-industrial establishments is threatening their integrity. Improved productivity of peri-urban agriculture can, therefore, make a very large contribution to meeting food security needs of cities as well as providing income to the peri-urban farmers. Hence, greenhouses agriculture can become an engine of pro-poor 'trickle-up' growth because of the synergistic effects of agricultural growth such as (UN, 2001):

- Increased productivity increases wealth.
- Intensification by small farmers raises the demand for wage labour more than by larger farmers.
- Intensification drives rural non-farm enterprise and employment.
- Alleviation of rural and peri-urban poverty is likely to have a knock-on decrease of urban poverty.

Despite arguments for continued large-scale collective schemes there is now an increasingly compelling argument in favour of individual technologies for the development of controlled greenhouses. The main points constituting this argument are summarised by (UN, 2001) as follows:

- Individual technologies enable the poorest of the poor to engage in intensified agricultural production and to reduce their vulnerability.
- Development is encouraged where it is needed most and reaches many more poor households more quickly and at a lower cost.
- Farmer-controlled greenhouses enable farmers to avoid the difficulties of joint management.

Such development brings the following challenges (UN, 2001):

- The need to provide farmers with ready access to these individual technologies, repair services and technical assistance.
- Access to markets with worthwhile commodity prices, so that sufficient profitability is realised.
- This type of technology could be a solution to food security problems. For example, in greenhouses, advances in biotechnology like the genetic engineering, tissue culture and market-aided selection have the potential to be applied for raising yields, reducing pesticide excesses and increasing the nutrient value of basic foods.

However, the overall goal is to improve the cities in accordance with the Brundtland Report (WCED, 1987) and the investigation into how urban green could be protected. Indeed, greenhouses can improve the urban environment in multitude of ways. They shape the character of the town and its neighbourhoods, provide places for outdoor recreation, and have important environmental functions such as mitigating the heat island effect, reduce surface water runoff, and creating habitats for wildlife. Following analysis of social, cultural and ecological values of urban green, six criteria in order to evaluate the role of green urban in towns and cities were prescribed (WCED, 1987). These are as follows:

- Recreation, everyday life and public health.
- Maintenance of biodiversity - preserving diversity within species, between species, ecosystems, and of landscape types in the surrounding countryside.
- City structure - as an important element of urban structure and urban life.
- Cultural identity - enhancing awareness of the history of the city and its cultural traditions.
- Environmental quality of the urban sites - improvement of the local climate, air quality and noise reduction.
- Biological solutions to technical problems in urban areas - establishing close links between technical infrastructure and green-spaces of a city.

The main reasons why it is vital for greenhouses planners and designers to develop a better understanding of greenhouses in high-density housing can be summarised as follows (WCED, 1987):

- Pressures to return to a higher density form of housing.
- The requirement to provide more sustainable food.
- The urgent need to regenerate the existing, and often decaying, houses built in the higher density, high-rise form, much of which is now suffering from technical problems.

The connection between technical change, economic policies and the environment is of primary importance as observed by most governments in developing countries, whose attempts to attain food self-sufficiency have led them to take the measures that provide incentives for adoption of the Green Revolution Technology (Herath, 1985). Since, the Green

Revolution Technologies were introduced in many countries actively supported by irrigation development, subsidised credit, fertiliser programmes, self-sufficiency was found to be not economically efficient and often adopted for political reasons creating excessive damage to natural resources. Also, many developing countries governments provided direct assistance to farmers to adopt soil conservation measures. They found that high costs of establishment and maintenance and the loss of land to hedgerows are the major constraints to adoption (Herath, 1985). The soil erosion problem in developing countries reveals that a dynamic view of the problem is necessary to ensure that the important elements of the problem are understood for any remedial measures to be undertaken. The policy environment has, in the past, encouraged unsustainable use of land (Herath, 1985). In many regions, government policies such as provision of credit facilities, subsidies, price support for certain crops, subsidies for erosion control and tariff protection, have exacerbated the erosion problem. This is because technological approaches to control soil erosion have often been promoted to the exclusion of other effective approaches. However, adoption of conservation measures and the return to conservation depend on the specific agro-ecological conditions, the technologies used and the prices of inputs and outputs of production.

13.1. Types of Greenhouses

Choosing a greenhouse and setting it up are important, and often expensive, steps to take. Greenhouses are either freestanding or lean-to, that is, built against an existing wall. A freestanding greenhouse can be placed in the open, and, hence, take advantage of receiving the full sun throughout the day. It is, therefore, suitable for a wide range of plants. However, its main disadvantage when compared to a lean-to type is that more heat is lost through its larger surface area. This is mainly why lean-to greenhouses have long been used in the walled gardens of large country houses to grow Lapageria rosea and other plants requiring cool, constant temperature, such as half-hardly ferns. However, generally, good ventilation and shading in the spring and summer to prevent overheating are essential for any greenhouse. The high daytime temperatures will warm the back wall, which acts as a heat battery, releasing its accumulated heat at night. Therefore, plants in a greenhouse with this orientation will need the most attention, as they will dry out rapidly.

Also, greenhouses vary considerably in their shapes and internal dimensions. Traditional greenhouses have straight sides, which allow the maximum use of internal space, and are ideal for climbers (Herath, 1985). On the other hand, greenhouses with sloping sides have the advantage of allowing the greatest penetration of sunlight, even during winter (Herath, 1985). The low winter sun striking the glass at 90°C lets in the maximum amount of light. Where the sun strikes the glass at a greater or lesser angle, a proportion of the light is reflected away from greenhouse. Sloping sides, also, offer less wind resistance than straight sides and therefore, less likely to be damaged during windy weather. This type of greenhouse is most suitable for short winter crops, such as early spring lettuce, and flowering annuals from seed, which do not require much headroom.

A typical greenhouse is shown schematically in Figure 10. However, there are several designs of greenhouses, based on dimensions, orientation and function. The following three options are the most widely used:

- A ready-made design
- A designed, which is constructed from a number of prefabricated modules
- A bespoke design

Of these, the prefabricated ready-made design, which is utilised to fit the site, is the cheapest greenhouses and gives flexibility. It is, also, the most popular option (Herath, 1985).

Specific examples of commercially available designs are numerous. Dutch light greenhouses, for example, have large panes of glass, which cast little shade on the plants inside. They are simple to erect, consisting of frames bolted together, which are supported on a steel framework for all but the smallest models. They are easy to move and extra sections can be added on to them, a useful attraction (Herath, 1985). Curvilinear greenhouses, on the other hand, are designed primarily to let in the maximum amount of light throughout the year by presenting at least one side perpendicular to the sun. This attractive style of greenhouse tends to be expensive because of the number of different angles, which require more engineering (Herath, 1985). Likewise, the uneven span greenhouses are designed for maximum light transmission on one side. These are generally taller than traditional greenhouses, making them suitable for tall, early season crops, such as cucumbers (Herath, 1985). Also, the polygonal greenhouses are designed more as garden features than as practical growing houses, and consequently, are expensive. Their internal space is somewhat limited and on smaller models over-heading can be a problem because of their small roof ventilations. They are suitable for growing smaller pot plants, such as pelargoniums and cacti (Herath, 1985). Another example is the solar greenhouses. These are designed primarily for areas with very cold winters and poor winter light. They take the form of lean-to structures facing the sun, are well insulated to conserve heat and are sometimes partially sunk into the ground. They are suitable for winter vegetable crops and early-sown bedding plants, such as begonias and pelargoniums (Herath, 1985). Mini lean-to greenhouses are suitable for small gardens where space is limited. They can, also, be used to create a separate environment within larger greenhouses. The space inside is large enough to grow two tomato or melon plants in growing bags, or can install shelves to provide a multi-layered growing environment, ideal for many small potted plants and raising summer bedding plants (Herath, 1985).

13.2. Construction Materials

Different materials are used for the different parts. However, wood and aluminium are the two most popular materials used for small greenhouses. Steel is used for larger structures and UPVC for conservatories (Jonathon, 1991).

13.3. Ground Radiation

Reflection of sunrays is mostly used for concentrating them onto reactors of solar power plants. Enhancing the insolation for other purposes has, so far, scarcely been used. Several years ago, application of this principle for increasing the ground irradiance in greenhouses, glass covered extensions in buildings, and for illuminating northward facing walls of buildings was proposed (Achard and Gicqquel, 1986). Application of reflection of sun's rays was motivated by the fact that ground illuminance/irradiance from direct sunlight is of very low intensity in winter months, even when skies are clear, due to the low incident angle of incoming radiation during most of the day. This is even more pronounced at greater latitudes. As can be seen in Figure 11, which depicts a sunbeam split into its vertical and horizontal components, nearly all of the radiation passes through a greenhouse during most of the day.

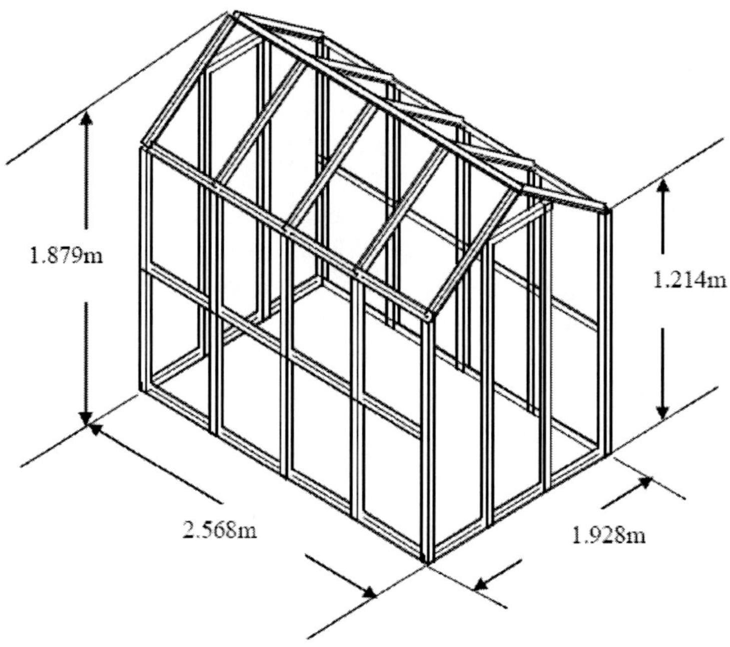

Figure 10. Greenhouse and base with horticultural glass.

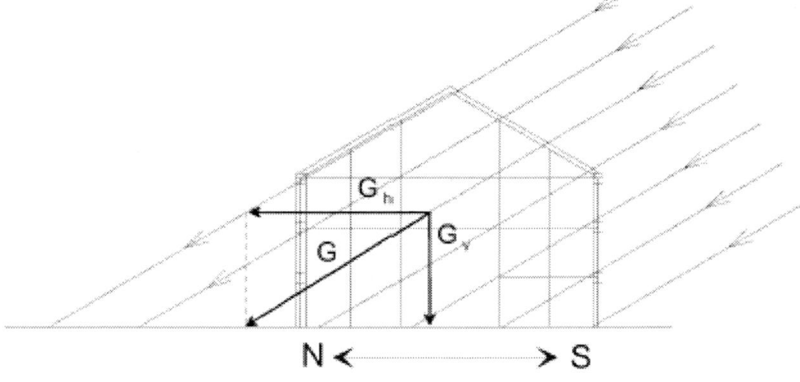

Figure 11. Relative horizontal and vertical components of solar radiation.

Large-scale, conventional, power plant such as hydropower, has an important part to play in development. It does not, however, provide a complete solution. There is an important complementary role for the greater use of small-scale, rural based, power plan (device). Such plant can be used to assist development since it can be made locally using local resources, enabling a rapid built-up in total equipment to be made without a corresponding and unacceptably large demand on central funds. Renewable resources are particularly suitable for providing the energy for such equipment and its use is also compatible with the long-term aims. It is possible with relatively simple flat plate solar collectors (Figure 12) to provide warmed water and enable some space heating for homes and offices which is particularly useful when the buildings are well insulated and thermal capacity sufficient for the carry over of energy from day to night is arranged. The perceptual reality arises from a multitude of sensorial components; visual, thermal, acoustic, olfactory and kinaesthetics.

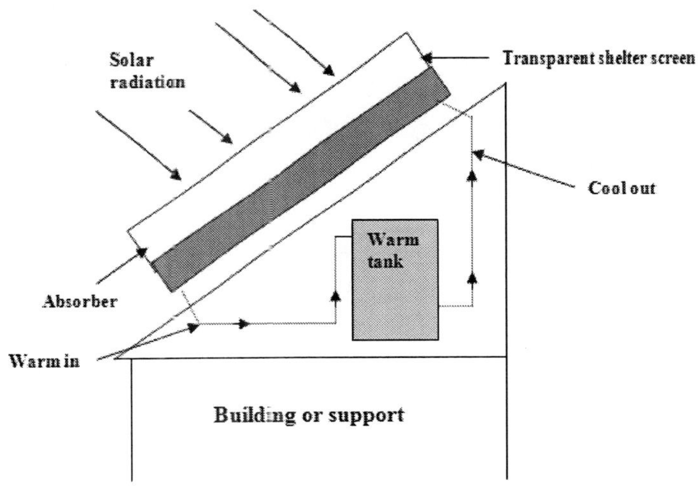

Figure 12. Solar heaters for hot water.

13.4. Greenhouse Environment

It has been known for long time now that urban centres have mean temperatures higher than their less developed surroundings. The urban heat increases the average and peak air temperatures, which, in turn, affect the demand for heating and cooling. Higher temperatures can be beneficial in the heating season, lowering fuel use, but they exacerbate the energy demand for cooling in summer time. In temperate climates neither heating nor cooling may dominate the fuel use in a building, and the balance of the effect of the heat is less. The solar gains, however, would affect the energy consumption. Therefore, lower or higher percentage of glazing, or incorporating of shading devices might affect the balance between annual heating and cooling load. As the provision of cooling is expensive with higher environmental cost, ways of using innovative alternative systems like mop fans will be appreciated (Figures 13-15). Indeed, considerable research activities have been devoted to the development of alternative methods of refrigeration and air-conditioning. The mop fan is a novel air-cleaning device that fulfils the functions of de-dusting of gas streams, and removal of gaseous contaminations from gas streams and gas circulation (Erlich, 1991). Hence, the mop fan seems particularly suitable for applications in industrial, agricultural and commercial buildings and greenhouses.

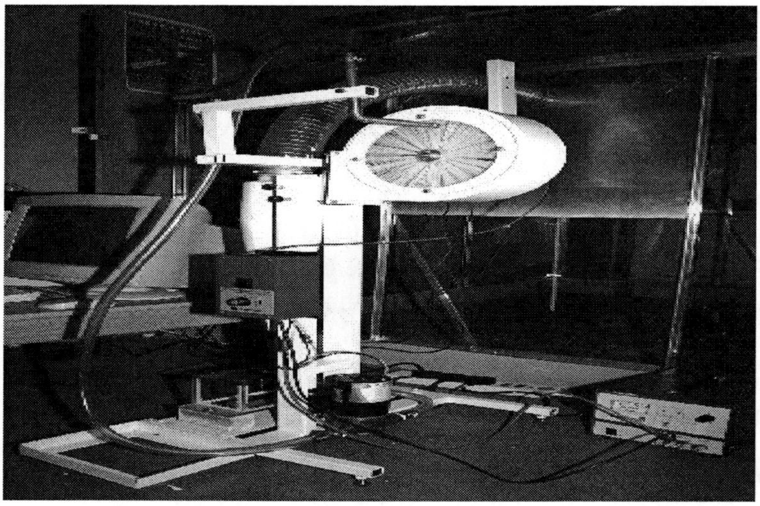

Figure 13. Mop fan systems.

Indoor conditions are usually fixed by comfort conditions, with air temperatures ranging from 15°C to 27°C, and relative humidities ranging from 50% to 70% (Abdeen, 2008). The system performance (COP) is defined as the ration between the cooling effect in the greenhouse and the total amount of air input to the mop fan. Hence,

COP = cooling delivered/air input to the mop fan (4)

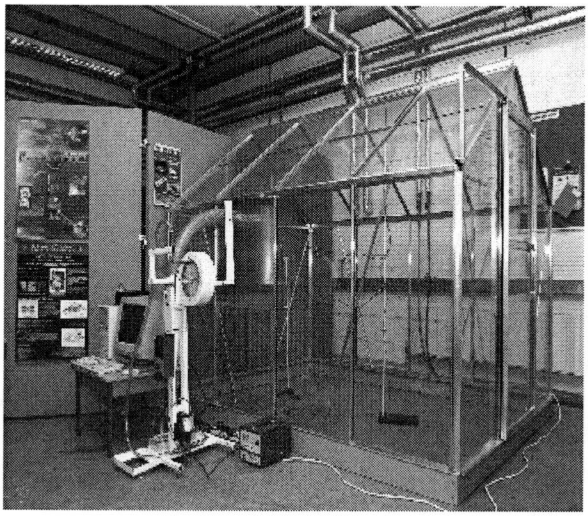

Figure 14. Mop fan in greenhouse.

Therefore, system performance (COP) varies with indoor and outdoor conditions. A lower ambient temperature and a lower ambient relative humidity lead to a higher COP. This means that the system will be, in principle, more efficient in colder and drier climates. The effect of indoor (greenhouse) conditions and outdoor (ambient) conditions (temperature and relative humidity) on system performance is illustrated in Figure 16. Air humidity is measured as a percentage of water vapour in the air on a scale from 0% to 100%, where 0% being dry and 100% being full saturation level. The main environmental control factor for dust mites is relative humidity.

Figure 15. Plants in greenhouse.

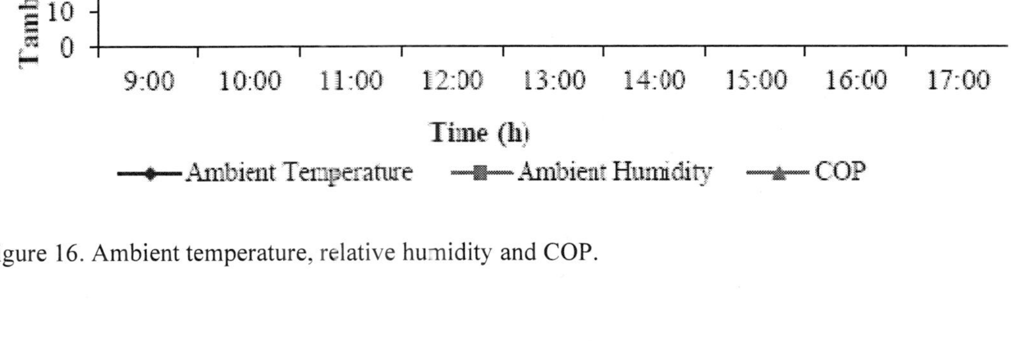

Figure 16. Ambient temperature, relative humidity and COP.

CONCLUSION

There is strong scientific evidence that the average temperature of the earth's surface is rising. This is a result of the increased concentration of carbon dioxide and other GHGs in the atmosphere as released by burning fossil fuels. This global warming will eventually lead to substantial changes in the world's climate, which will, in turn, have a major impact on human life and the built environment. Therefore, effort has to be made to reduce fossil energy use and to promote green energies, particularly in the building sector. Energy use reductions can be achieved by minimising the energy demand, by rational energy use, by recovering heat and the use of more green energies. This study was a step towards achieving that goal. The adoption of green or sustainable approaches to the way in which society is run is seen as an important strategy in finding a solution to the energy problem. The key factors to reducing and controlling CO_2, which is the major contributor to global warming, are the use of alternative approaches to energy generation and the exploration of how these alternatives are used today and may be used in the future as green energy sources. Even with modest assumptions about the availability of land, comprehensive fuel-wood farming programmes offer significant energy, economic and environmental benefits. These benefits would be dispersed in rural areas where they are greatly needed and can serve as linkages for further rural economic development. The nations as a whole would benefit from savings in foreign exchange, improved energy security, and socio-economic improvements. With a nine-fold increase in forest – plantation cover, a nation's resource base would be greatly improved. The international community would benefit from pollution reduction, climate mitigation, and the increased trading opportunities that arise from new income sources. The non-technical issues, which have recently gained attention, include: (1) Environmental and ecological factors, e.g., carbon sequestration, reforestation and revegetation. (2) Renewables as a CO_2 neutral replacement for fossil fuels. (3) Greater recognition of the importance of renewable energy, particularly modern biomass energy carriers, at the policy and planning levels. (4) Greater recognition of the difficulties of gathering good and reliable renewable energy data, and efforts to improve it. (5) Studies on the detrimental health efforts of biomass energy particularly from traditional energy users. Two of the most essential natural resources for all life on the earth and for man's survival are sunlight and water. Sunlight is the driving force behind many of the renewable energy technologies. The worldwide potential for utilising this resource, both directly by means of the solar technologies and indirectly by means of biofuels, wind and hydro technologies is vast. During the last decade interest has been refocused on renewable energy sources due to the increasing prices and fore-seeable exhaustion of presently used commercial energy sources. Plants, like human beings, need tender loving care in the form of optimum settings of light, sunshine, nourishment, and water. Hence, the control of sunlight, air humidity and temperatures in greenhouses are the key to successful greenhouse gardening. The mop fan is a simple and novel air humidifier; which is capable of removing particulate and gaseous pollutants while providing ventilation. It is a device ideally suited to greenhouse applications, which require robustness, low cost, minimum maintenance and high efficiency. A device meeting these requirements is not yet available to the farming community. Hence, implementing mop fans aides sustainable development through using a clean, environmentally friendly device that decreases load in the greenhouse and reduces energy consumption. It is found that densification of towns could have both positive and negative effects on the total energy demand. With suitable urban and building design details, population should and could be accommodated with minimum worsening of the environmental quality.

References

Abdeen, M. O. (2008). Chapter 10: Development of integrated bioenergy for improvement of quality of life of poor people in developing countries, In: Energy in Europe: Economics, Policy and Strategy- IB, Editors: Flip L. Magnusson and Oscar W. Bengtsson, 2008 NOVA Science Publishers, Inc., p.341-373, New York, USA.

Achard, P., and Gicqquel, R. (1986). European passive solar handbook. Brussels: Commission of the European Communities.

Bos, E., My, T., Vu, E., and Bulatac R. (1994). World population projection: 1994-95. Edition, published for the World Bank by the John Hopkins University Press. Baltimore and London.

Brain, G., and Mark, S. (2007). Garbage in, energy out: landfill gas opportunities for CHP projects. *Cogeneration and On-Site Power* 8 (5): 37-45.

Erlich, P. (1991). *Forward facing up to climate change, in Global Climate Change and Life on Earth*. R.C. Wyman (Ed), Chapman and Hall, London.

Herath, G. (1985). The green revolution in Asia: productivity, employment and the role of policies. *Oxford Agrarian Studies*. 14: 52-71.

International Energy Agency (IEA). (2007). Indicators for Industrial Energy Efficiency and CO_2 Emissions: *A Technology Perspective*.

John, W. (1993). The glasshouse garden. The Royal Horticultural Society Collection. UK.

Jonathon, E. (1991). Greenhouse gardening. The Crowood Press Ltd. UK.

Omer, A. M. (2009a). Environmental and socio-economic aspect of possible development in renewable energy use, In: Proceedings of the 4th International Symposium on Environment, Athens, Greece, 21-24 May 2009.

Omer, A. M. (2009b). Energy use, environment and sustainable development, In: Proceedings of the 3rd International Conference on Sustainable Energy and Environmental Protection (SEEP 2009), Paper No.1011, Dublin, Republic of Ireland, 12-15 August 2009.

Omer, A. M. (2009c). Energy use and environmental: impacts: a general review, Journal of Renewable and Sustainable Energy, Vol.1, No.053101, p.1-29, United State of America, September 2009.

Omer, A. M. (2009d). Chapter 3: Energy use, environment and sustainable development, In: Environmental Cost Management, Editors: Randi Taylor Mancuso, 2009 NOVA Science Publishers, Inc., p.129-166, New York, USA.

Omer, A.M., and Yemen, D. (2001). Biogas an appropriate technology. Proceedings of the 7th Arab International Solar Energy Conference, P.417, Sharjah, UAE, 19-22 February 2001.

Oxburgh, E.R. (1975). Geothermal energy. *Aspects of Energy Conversion*. p. 385-403.

Rawlings, R.H.D. (1999). Technical Note TN 18/99 – Ground Source Heat Pumps: A Technology Review. Bracknell. The Building Services Research and Information Association. Robinson, G. (2007) Changes in construction waste management. *Waste Management World* p. 43-49. May-June 2007.

Sims, R.H. (2007). Not too late: IPCC identifies renewable energy as a key measure to limit climate change. *Renewable Energy World* 10 (4): 31-39.

Swift-Hook, D.T., et al. (1975). Characteristics of a rocking wave power devices. *Nature* 254: 504.

Trevor, T. (2007). Fridge recycling: bringing agents in from the cold. *Waste Management World* 5: 43-47.

United Nations (UN). (2001). World Urbanisation Prospect: The 1999 Revision. New York. The United Nations Population Division.

WCED. (1987). Our common future. New York. Oxford University Press.

ABBREVIATIONS

a	annum
ha	hectares
l	litre

In: Air Pollution Emissions
Editors: Daniel G. Vasiliev and Robert A. Kask
ISBN: 978-1-62100-453-0

Chapter 2

HYDROCARBON AIR POLLUTION LASER MONITORING

Valery G. Shemanin*
Kuban State Technological University, Novorossiysk Polytechnic Institute, Novorossiysk, Russia

ABSTRACT

The hydrocarbons molecules of the different composition are the most pollutants in the atmosphere under the oil - gas converting or distributing enterprises. These molecules in atmospheric air monitoring can be realized by the laser remote sensing systems using. This allows to measure the pollutants concentration at the large sensing distances up to tenth of kilometers with the high repetition rate or time resolution. The experimental studies and computer simulation results were stated that the differential absorption and scattering lidar and Raman lidar have the preference possibilities in this problem solving as have been seen from our previous papers.

1. THE HYDROCARBON MOLECULES

The hydrocarbon molecules are the most important pollutants in the atmosphere under the enterprises of the oil-gas industrial. The oil, natural gas and their converting products that widely used as the fuel and raw material for some chemical products creation are consists of the hydrocarbons. All of the hydrocarbons molecules classes including paraffin, benzenes and sulphur-containing hydrocarbons are the most spread pollutants in the atmospheric air. And this hydrocarbons pollution is the global biosphere pollution.

The hydrocarbon molecules are exposed to any conversion in the atmospheric air: oxidation, polymerization and other chemical reaction interacting with the other atmospheric pollutants under the Sun radiation excitation. The peroxide compounds, free radicals, hydrocarbons compounds with the nitrogen and sulphur oxides were formed often as the

* E-mail: vshemanin@nbkstu.org.ru.

aerosol particles in these chemical reaction results. And these reactions products are the very toxic pollutants. They concentrate under the inversion layer – the cold air is under the hot one - and hydrocarbons consistence near the earth arises sharply that becomes one of the origins unknown earlier in the nature photochemical fog. The paraffin determine the power and density of the this photochemical fog thus they are not own the dangerous pollutants.

The pollutants limit low permissible level (LPL) in the atmospheric air has been stated [1] and this level have not to arise in agreement to the law demands. The pollutants molecules parameters and its LPL are exhibited in Table 1.

The oil and oil products transporting process requires to know in mind the environmental problems of this process. And the oil consists of some hydrocarbons and its monitoring in atmosphere under the industrial enterprise can be realized by the hydrocarbons molecules laser remote sensing. Some years ago pollutants emission from the oil emission source controlling was lead to sum determination of the hydrocarbon consistence and its maximal one time LPL is consists of 5 mg/m^3 [2]. Thus such an approach has not reflected the real layout because the emission chemical composition was not taken into account.

The pollutions safety level for paraffin two group C_1—C_5 and C_6—C_{10}, that are equal to 50 and 30 mg/m^3 accordingly [2] have been determined due to specifying the oil emission hydrocarbon composition and pollutants influence character on the human health. Besides the more toxic pollutants which ratio is the small part in the mixture monitoring was directed to lead with the paraffin that were formed the main ratio in the emission. In particularly hydrocarbons emission from oil source can include the methantiol and xylol molecules that LPL are some power down about the paraffin LPL and are equal to 0,8 mg/m^3 and 0,2 mg/m^3 accordingly [1].

Table 1. The shot characteristics of paraffin molecules

Pollutant	**LPL, mg/m^3**	**Mole mass, g/mole**	**Molecule CH stretching wave number ν_0^{**}, cm^{-1}**	**LPL, cm^{-3}**
Methane (CH_4)	50	16	2914	$2 \cdot 10^{15}$
Ethane (C_2H_6)	50	30	2954	10^{15}
Butane (C_4H_{10})	200	58	2890	$2.1 \cdot 10^{15}$
Pentane (C_5H_{12})	100	72	2885	$8.4 \cdot 10^{14}$
Hexane (C_6H_{14})	60	86	2886	$4.2 \cdot 10^{14}$

Thus this problem requires new laser methods and systems for the atmospheric air pollution monitoring. And the next description is the hydrocarbons laser remote sensing studies for this problem solution.

2 The Differential Absorption and Scattering Lidar Experimental Studies and Computer Simulation Results

The resonant absorption cross section in the visible and infrared spectrum regions exceeds as usually considerable as the effective fluorescence cross section with the quenching and Raman cross section [3]. Thus the resonant absorption cross section in visible and infrared spectrum exceeds as usually magnificently as the effective fluorescence cross section with the quenching and Raman cross section [3]. The sensing method for the chosen molecular component aversive concentration level along the distance can be created using the laser radiation extinction with the corresponding wavelength. The differential absorption method was usually applied to distinguish the studied molecule absorption contribution into the laser beam extinction [4]. It is necessary the two laser beams using in the differential absorption and scattering lidar which extinctions were determined by the back scattering signals. One beam lies in this molecule absorption band maximum and other – over this band. The studied molecules concentration distribution in the atmosphere data was derived from the recorded signals comparison at these two wavelengths in strongly weak spectral interval [3]. This determines the experimental instruments realized this method main distinguishes – the measurements limit high sensitivity in the real time scale and distance resolution at the remote sensing in the large distance ranging. This method has been suggested at first for the water vapor consistence remote sensing in the atmosphere in [3]. The space resolution and power signals at the two laser radiation wavelengths were responsible by the molecules absorption high cross section and the signals ratio gives the differential absorption value. And it gives the possibility for the sensing at the large ranging distance. The new infrared photodetectors have the high sensitivity and the differential absorption becomes the more universal method [5].

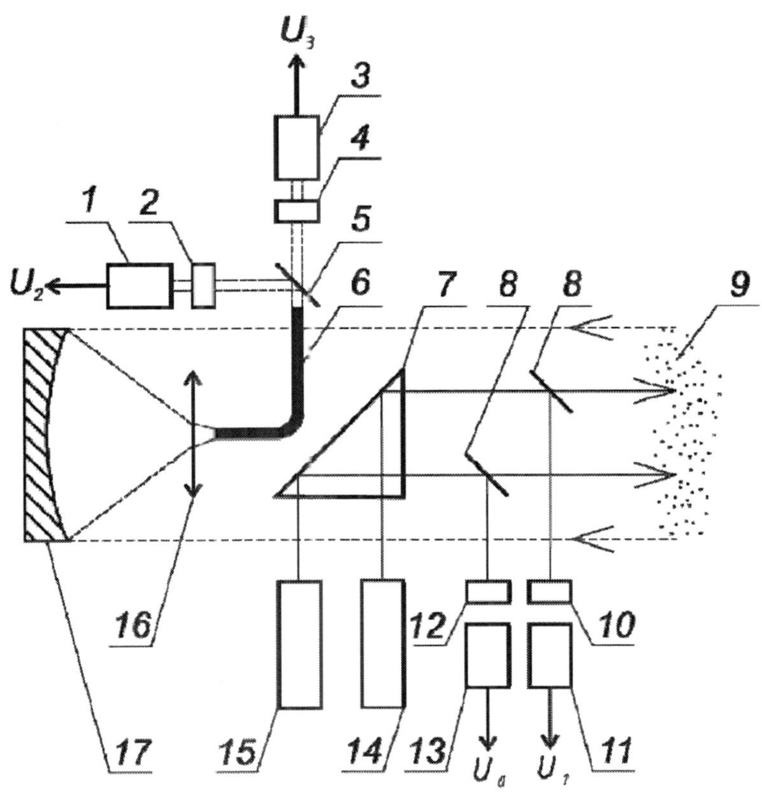

Figure 1. The differential absorption and scattering lidar optical layout: 1, 3 – photodetector; 2, 4, 10, 12 – interference filters; 5 – mirror with reflectance about 0.5; 6 – light waveguide fiber; 7 – prism; 8 – glass plate; 9 – target; 11, 13 – photodiode; 14, 15 – laser; 16 – lens objective and 17 – spherical mirror.

Such a differential absorption and scattering lidar variant optical layout from [6] was shown in Figure 1 for its operation description. The lidar was designed by coaxial layout [3, 5] when the two beams have parallel directions along the receiving telescope axis. The laser 15 radiation was directed into the atmosphere along this axis. Its intensity at the sensing wavelength passes the atmosphere forward and back was weaken by the studied molecules absorption and directed through the interference filter 4 to the photodetedtor 3. Simultaneously the laser 14 radiation was reflected by the mirror 5 through the interference filter 2 to the photodetedtor 1. These photodetectos were served for the two lasers radiation intensities. The second laser radiation passed the atmospheric layer two times and it is the reference channel in the differential measurement scheme. The two photodetectors 1 and 3 output pulses were input to the PC controlling measuring system. And two photodiodes 11 and 13 were served for the laser pulse energy measuring and all of the lidar operation synchronization.

It is interesting task to estimate the such a lidar potential possibilities in the studied molecules concentration range as a ranging distance function. The lidar equation for the differential absorption and scattering analysis can give the answer to this question. And paraffin molecules concentration remote measurements in the atmosphere process can be analyzed with this lidar equation that can be derived as earlier in [4, 6].

The lidar equation for the elastic back scattering was wrought according to [3, 6] in form

$$P(\lambda,R)=P_L(\lambda_L)K_1A_0T^2(\lambda_L,R)\rho/R^2 \tag{1}$$

where $P(\lambda,R)$ – back scattering signal power at the photodetector from the distance R at the wavelength λ_L;

$P_L(\lambda_L)$ – the laser radiation power and its wavelength;

K_1 – the lidar constant;

A_0 – the receiving telescope cross section square;

ρ - the topographic target reflection coefficient or Mi elastic scattering and Raleigh molecular scattering sum coefficient.

The atmospheric transmissions at the two laser radiation wavelengths are equal:

$$T(\lambda_L, R) = \exp\left[-\int_0^R k(\lambda_L, R)dR\right] \text{ и } T(\lambda, R) = \exp\left[-\int_0^R k(\lambda, R)dR\right] \tag{2}$$

where $k(\lambda_L, R)$ и $k(\lambda, R)$ - the extinction coefficients in the atmosphere for the laser and recorded wavelengths.

The studied molecules concentration value is consists in the exponent power $k(\lambda_L,R)$ that is determined in general case by the equation [3, 4]

$$k(\lambda_L, R) = k_A(\lambda_L, R) + N(R)\sigma_0(\lambda_L) \tag{3}$$

where the first adding in the right part is the atmospheric extinction coefficient at the laser radiation wavelength and the second – the multiplication of the studied molecules concentration value and these molecules resonant absorption cross section.

Following the differential absorption and scattering idea [4] the two lidar equation (1) for two laser radiations wavelengths λ_0 and λ_1 with the standing formula (2) were taken and divided the first to the second one. As the dividing result the lidar equation for the general case of the differential absorption and scattering was received in the suggestion about the difference all of the wavelengths depending multipliers:

$$\frac{P(\lambda_0,R)}{P(\lambda_1,R)}=\frac{P_0K_{10}\rho_0}{P_1K_{11}\rho_1}\exp\{-2\int_0^R[k(\lambda_0,R)-k(\lambda_1,R)]dR\} \tag{4}$$

The correlation for the extinction coefficients at the two wavelengths (3) were put in formula (4) and the lidar equation for the differential absorption and scattering was rewritten at the end in form

$$\int_0^R N(R)dR=\frac{1}{2\sigma_0}\ln[\frac{P(\lambda_1,R)P_0K_{10}\rho_0}{P(\lambda_0,R)P_1K_{11}\rho_1}]-\int_0^R[k_A(\lambda_0,R)-k_A(\lambda_1,R)]dR \tag{5}$$

Thus this lidar equation for the differential absorption and scattering gives the possibility the optical densities simulation for the studied paraffin molecules chosen concentration range and wavelengths in the chosen ranging distances for the lidar system optimal variant seeking.

And now let us fulfill this lidar equation (5) simulation as in [4, 6]. The studied paraffin molecule infrared absorption band maximum wavelength is equal to 3.4 mkm [3, 7]. The laser radiation wavelength out of this band have been taken 2.01 mkm that lies in the atmospheric transparency region. Such a lidar experimental layout can be realized by the solid pulse laser as in [8]. The other magnitudes in lidar equation are the next:

- $A_0 = 0.008$ m ^{2}; $K_1 =$ 0,4 (experimental result for the 1.06 mkm wavelength [5]);
- the laser radiation peak power P_L= 1, 10 and 100 kW and the two laser radiation power ratio at the two chosen wavelengths is equal to the reverse ratio of the photodetectors spectral sensitivity values at these wavelengths;
- ranging distance $R = 0.1, 0.5 \ldots 5$ km with 100 m space resolution;
- the studied molecules concentration range $10^{10} \ldots 10^{16}$ cm $^{-3}$;
- the avalanche photodiode spectral sensitivity values at the chosen wavelengths were taken from [9] and equal to 0.1 and 0.05 of the maximal value at the 1.4 mkm wavelength accordingly;
- the atmospheric extinction coefficient values k_A were resulted in [9] and were equal to 0.014 и 0.0105 km^{-1} accordingly for these wavelengths;

- the paraffin molecules resonant absorption cross section values were appreciated by the data from [3] and the hexane molecules experimental lidar studies additional data give the value of $\sigma_0 = 8\ 10^{-19}$ cm ^{2};
- the topographic target reflectance coefficient values were estimated by the data [5] and were taken for rectangle reflector - 0.3 and the mat plane – 0.15. The back scattering coefficient in the atmosphere sum value is equal to 10^{-7} as in [3].

The differential absorption optical density values simulation has been fulfilled by the lidar equation (5) using the above mentioned data for the chosen studied molecules concentrations and laser radiation wavelengths in the ranging distance from 0.1 up to 5 km. The plot of the optical density logarithm *lgD* versus ranging distance *R* dependence is exhibited in Figure 2.

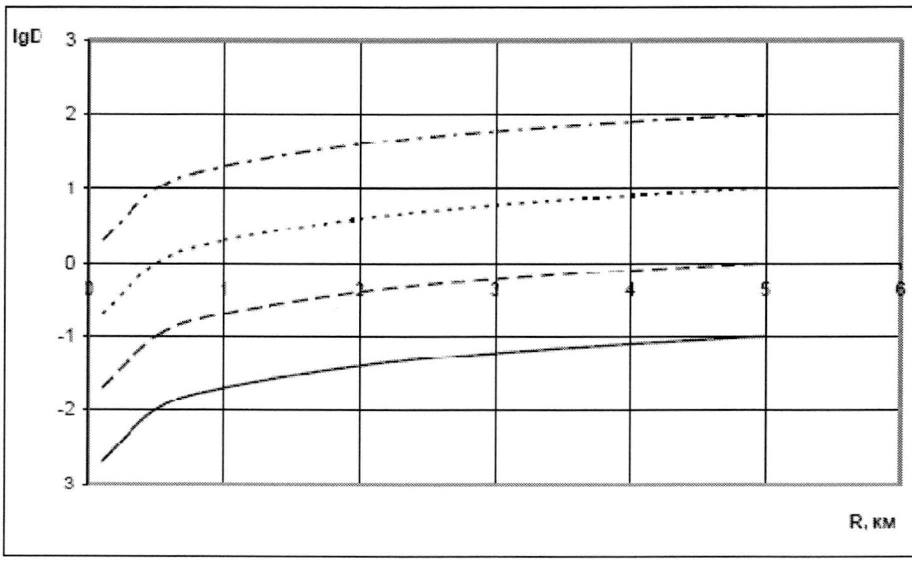

Figure 2. The plot of the optical density logarithm lg D versus ranging distance R for the different values of the paraffin molecules concentration level of 10^{10}, 10^{12}, 10^{14}, 10^{16} cm^{-3} in our experimental condition.

It can be seen that this optical density *D* value has the minimum for the low concentration levels and distances but maximal value - for the high concentration levels and any distances. The simulation was not made for the optical density values more than 10 because the avalanche photodiode dynamic range has been chosen as 10^4 in accordance with [3, 9].

It is need to mark that the possible concentrations range 10^{12} - 10^{16} cm^{-3} for the distance range 10 m and the integral value along the distance 5 km - 10^8 - 10^{13} cm^{-3}. And it can measure the paraffin molecules concentration in the range 10^{12} - 10^{16} cm^{-3} with the space resolution of 100 m along the ranging distance up to 5 km by such a lidar. The plot of the logarithm optical density versus paraffin concentration level dependence is shown in Figure 3 at the distances of 1 and 5 km.

It is interesting to calculate the lidar signal power in equation (1) for the different experimental situation and to compare these values with the Sun background power values that evaluated by the background spectral brightness in the infrared spectrum $S_b(\lambda)$ from [3]. These background power values $P_b(\lambda, R)$ for our experimental layout with the $S_b(\lambda)$ values were calculated by the equation from [10].

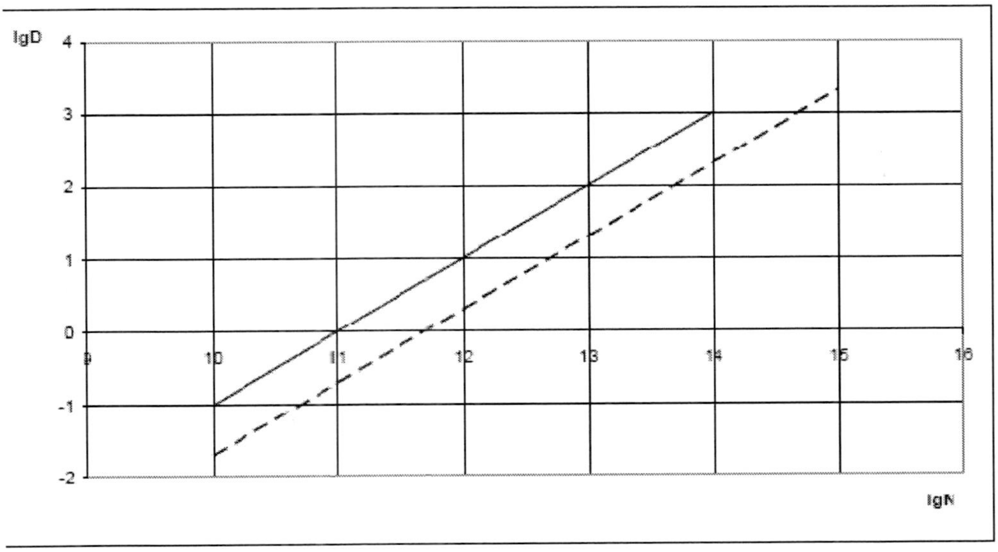

Figure 3. The plot of the optical density logarithm *lg D* versus the paraffin molecules concentration level logarithm *lg N* at the ranging distance *R* values of 1 and 5 km.

$$P_b(\lambda, R) = S_b(\lambda)T(\lambda, R)K_2\xi_p(\lambda)A_0\Omega(R)\Delta\lambda \quad (6)$$

where $\Omega(R)$ - the receiving telescope field of view space angle and it is equal to $\Omega(R)$ $=A_0/R^2$; but A_0 - the receiving telescope cross section, $\Delta\lambda$ - the receiving system spectral bandwidth.

The Sun background power value is equal to 80·pW for this experimental condition. The calculated results are exhibited in Figure 4. The plot of the elastic back scattering power versus ranging distance dependence and the background power at the laser radiation wavelengths of 3.4 mkm are shown in Figure 4.

The paraffin molecules can be detected in all of the distance range as it follows from Figure 4. The more long distance lidar sensing variant consideration notes that the background power begins to exceed the back scattering signal at the distance more than 35 km. The lidar signal power decreases with the ranging distance magnification that leads to the impossibility of the high concentration laser sensing at the large distance. These results analysis shows that lidar signal power range exceeds some power and such a lidar can not record all of the power range. The laser radiation with the changing power is the optimal variant for such a system and laser radiation power can be changed with the studied paraffin concentration that allows to record the back scattering power maximal value in the distance range 0.1 … 5 km.

Therefore these results allows the possibility of the laser radiation parameters optimal choosing for the paraffin concentration sensing in the atmosphere by the differential absorption and scattering lidar at the required distance and concentration in the night and daytime condition.

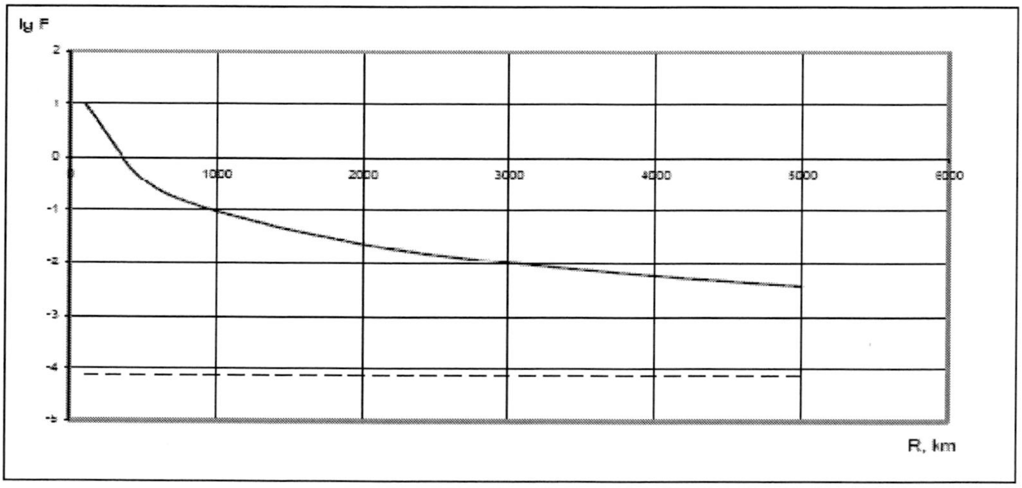

Figure 4. The plot of the elastic back scattering power (in mW) logarithm *lg P* versus ranging distance *R* at the wavelength of 3,4 mkm in comparison with the Sun background power logarithm as the line parallel to the abscissa axis in our experimental condition.

3. Raman Lidar Experimental Studies and Computer Simulation Results

Lidar systems for the pollution molecules concentration remote measurements in the atmosphere can become a basis for the atmospheric air quality monitoring systems and the emergency pollution emissions prediction in the atmosphere under industrial enterprises [3, 5, 8, 11].

The lidar systems parameters detailed studies for each experimental situation are required for such controlling systems designing. Besides, for each type of the gaseous pollutant substance low limit permissible concentration level (LPL) in atmospheric air [1] are exhibited in Table 1. The laser system variants for the environmental monitoring problems solving and the atmospheric pollution control have been suggested earlier in [8, 11].

The Raman lidar system for the paraffin molecules concentration level in the gaseous emissions into the atmosphere under industrial enterprise will be considered in this part of studies. Therefore the present work goal is the Raman lidar potential possibilities estimation for the hydrocarbons molecules sensing in atmosphere for such molecules concentration level detection in atmosphere over industrial region [11].

Raman lidar application for the operative space area sensing and the hydrocarbons molecules concentration distribution measurement was proved to be perspective according to the results in [12, 13] for such lidar sensing potential possibilities for the hydrogen and hydrogen fluoride molecules in atmosphere.

The Raman scattering lidar optical layout for the studied hydrocarbon molecules is remained practically the same as in [5] and is presented in Figure 5.

The lidar optical module consists of the laser 10, the photodiode 9 for the laser impulse energy controlling and the lidar operation synchronization (signal U_0), the Newton type receiving telescope with the spherical mirror 1 of 0.4 m diameter and the photo detector including the lens objective 4, interferential filter 3 with the 628 … 631 nm Raman wavelength maximum and 5 nm half width and FEU-79 2 type photomultiplier tube (PMT). This FEU signal U_1 is proportional to the Raman pulse energy for the hydrocarbon molecules.

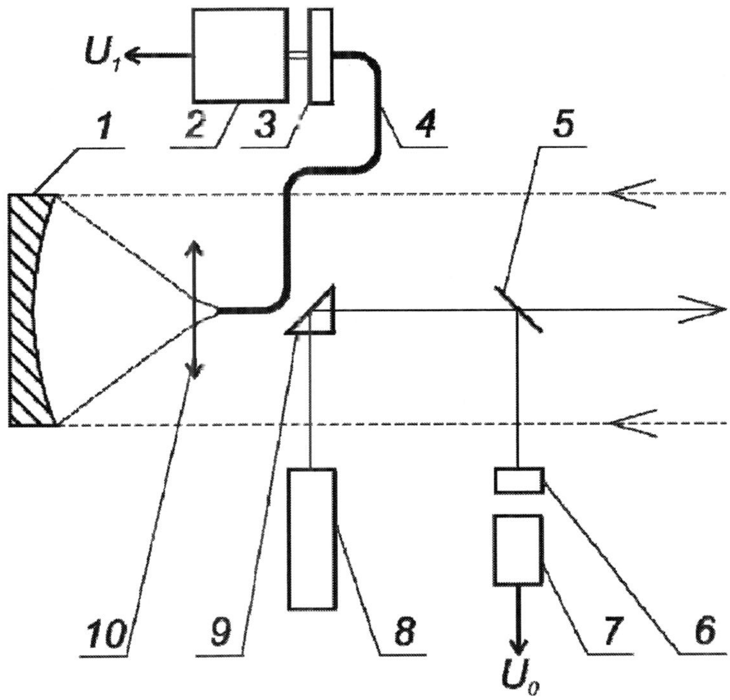

Figure 5. The Raman lidar optical layout.

1- spherical mirror, 2 - PM, 3, 6 – interferential filter, 4 – light waveguide, 5 – glass plate, 7 – photodiode, 8 – laser, 9 - block of prisms, 10 - lens objective.

The laser radiation part was used for the reference signal creation which sets time reference mark and its amplitude – laser pulse energy. The Raman back scattering from this volume was gathered by the Newton type telescope to the FEU photocathode. The FEU output pulse was connected to the microprocessor measuring system input operation with the personal computer and it was processed by LIDAR specialized codes.

For the lidar system designing the laser radiation optimum wavelength choice is necessary for the such a lidar photo detector recording the maximum Raman pulse energy of the hydrocarbon molecules in the daytime sensing conditions.

The studied molecules recorded Raman emission power can be described by the Raman lidar equation in our experimental situation. This Raman back scattering lidar equation by the studied molecules will be considered in a form [5, 12]:

$$P(\lambda, R) = P_L K_1 \Delta R A_0 T(\lambda_L, R) T(\lambda, R) (\frac{d\sigma}{d\Omega}) N / R^2 \qquad (7)$$

where $P(\lambda, R)$ - the Raman signal power at the photo detector and wave length λ, sensing from the distance R;

P_L - the laser radiation power and λ_L - its wave length;

K_1 – the lidar constant [13];

$\Delta R = \frac{c\tau_L}{2}$ - the ranging distance step,

c – the light velocity, and

τ_L –one measurement time duration, and its minimum value is equal to the laser radiation pulse duration;

A_0 - the receiving telescope cross section;

$T(\lambda_L,R)$, $T(\lambda,R)$ - the atmospheric transmission at the wave lengths of the laser radiation and the Raman signal accordingly;

$(d\sigma / d\Omega)$ - the Raman differential cross by the studied molecules at the laser radiation wave length and

N – the studied molecules concentration.

For the such a lidar potential possibilities estimation the lidar equations (7) computer simulation have been fulfilled for the hydrocarbon molecules vibration Raman back scattering with the concentration level of 10^{19} cm^{-3} in the ranging distance up to 2000 m.

The YAG-Nd lasers radiation of the 532, 355 and 266 nm wavelengths have been chosen for the sensing as earlier in [5]. The paraffin molecules Raman C- H stretching band wavelengths have been calculated for these wavelengths with the wave number values from Table 2 [3, 7].

The studied molecules Raman band back scattering differential cross sections have been calculated, as in [5], by the experimental cross section values for the 337 nm laser radiation wavelength and the chosen Raman bands from [8] and are also presented in Table 2. The FEU photocathode spectral sensitivity values $\xi(\lambda)$ for the studied molecules Raman bands wavelengths are also presented in Table 3. The atmospheric transmission along the ranging distance values have been calculated as earlier in [5] by the atmospheric extinction coefficient values k from [5, 9] and are collected in Table 3. The other parameters values in the lidar equation (7) were the next: the ranging distance step is equal to 7,5 m for the measurement time duration of 50 ns, lidar constant K_2 = 0,495, the receiving telescope square A_0 = 0,1256 m^2

The studied molecules Raman power numerical calculations have been fulfilled with these data at the concentration level of 10^{19} cm^{-3} in the chosen ranging distances from 100 up to 2000 m for all laser radiation wavelengths with peak power of 100 kW and the laser pulses repetition rate values 10 Hz. The plot of the ethane molecules Raman power logarithm versus the ranging distance R as an example for all the laser radiation wavelengths is resulted in Figure 5. The background power has been calculated as in [5] with the Sun spectral brightness data from Table 3.

Table 2. The laser radiation and Raman bands wavelength, Raman differential cross section for the studied molecules values

Molecule	Methane CH_4		Ethane C_2H_6		Propane C_3H_8	
$\tilde{V}$, cm^{-1}	2916		2954		2886	
λ_L, nm	$\lambda_{\kappa p}$, nm	$(d\sigma/d\Omega)\cdot10^{30}$, cm^2/sr	$\lambda_{\kappa p}$, nm	$(d\sigma/d\Omega)\cdot10^{30}$, cm^2/sr	$\lambda_{\kappa p}$, nm	$(d\sigma/d\Omega)\cdot10^{30}$, cm^2/sr
532	629,6	2,8	628,5	7,5	631,2	13,2
266	288,4	44,5	288,1	126,8	288,7	210,4
355	396,0	14,0	395,5	38,1	396,6	66,3
308	338,4	24,8	338,0	67,2	338,8	117,1
Molecule	**Butane C_4H_{10}**		**Pentane C_5H_{12}**		**Hexane C_6H_{14}**	
$\tilde{V}$, cm^{-1}	2890		2885		2886	
λ_L, nm	$\lambda_{\kappa p}$, nm	$(d\sigma/d\Omega)\cdot10^{30}$, cm^2/sr	$\lambda_{\kappa p}$, nm	$(d\sigma/d\Omega)\cdot10^{30}$, cm^2/sr	$\lambda_{\kappa p}$, nm	$(d\sigma/d\Omega)\cdot10^{30}$, cm^2/sr
532	628,7	15,0	628,5	19,9	628,5	21,5
266	288,2	239,7	288,1	319,1	288,1	344,7
355	395,6	75,6	395,5	100,6	395,5	102,7
308	338,1	133,4	338,0	177,5	338,0	191,8

Table 3. The laser radiation and Raman bands wavelength for the studied molecules, atmospheric extinction coefficient, photodetector spectral sensitivity and the Sun light spectral brightness values

	λ, nm	$k(\lambda,R)$, km^{-1}	$\xi_p(\lambda)$	$S_b(\lambda)$, W/m^2 sr nm
Laser	532	0,16	-	-
	266	0,785	-	-
	355	0,31	-	-
	308	0,45	-	-
CH_4	629,6	0,15	0,48	$13,2\cdot10^{-1}$
	288,4	0,54	0,36	$8,0\cdot10^{-5}$
	396,0	0,24	0,88	$10,6\cdot10^{-3}$
	338,4	0,36	0,42	$9,5\cdot10^{-4}$
C_2H_6	631,2	0,15	0,48	$13,2\cdot10^{-1}$
	288,7	0,54	0,36	$8,0\cdot10^{-5}$
	396,6	0,24	0,88	$10,6\cdot10^{-3}$
	338,8	0,35	0,39	$9,5\cdot10^{-4}$
C_3H_8, C_4H_{10},	628,5	0,15	0,48	$13,2\cdot10^{-1}$
C_5H_{12}, C_6H_{14}	288,1	0,55	0,25	$8,0\cdot10^{-5}$
	395,5	0,25	0,83	$10,6\cdot10^{-3}$
	338,0	0,36	0,40	$9,5\cdot10^{-4}$

The Raman power values for other studied molecules and our experimental situation are even more due to the great values of these molecules Raman differential cross section as it follows from Table 2. The best result is got for the laser radiation 266 nm wavelength as it can be seen from Figure5. But it is better to use the 355 nm wavelength laser radiation from the strong UV spectrum radiation extinction in the atmosphere/ Therefore the best results for all the molecules and this 355 nm wavelength are collected in Figure 6.

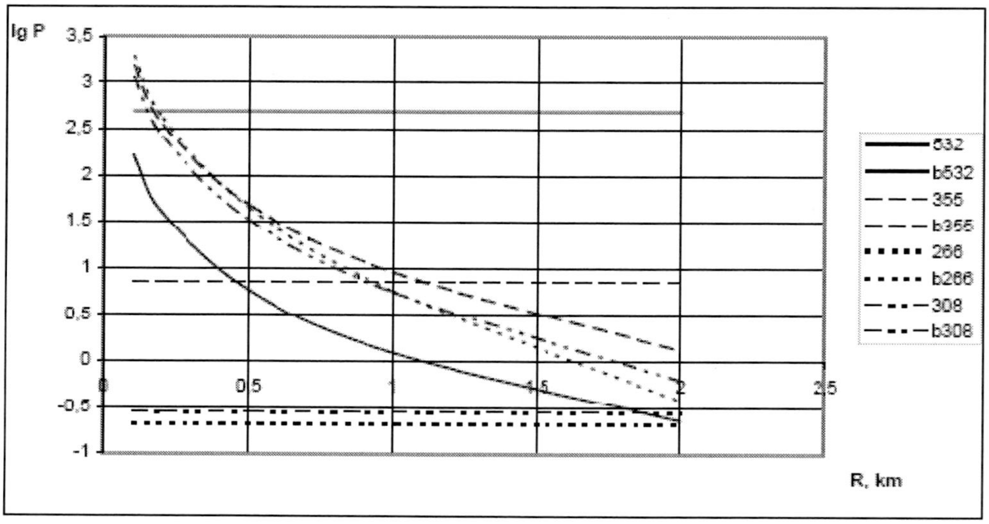

Figure 6. The plot of the ethane molecules Raman power values logarithm *lgP* (in nW) versus ranging distance *R* for all laser radiation wavelengths and molecules concentration of f 10^{19} cm^{-3}. The lines parallel to the abscissa axis b is the Sun background power level.

These results do not reflect all of the lidar system for the hydrocarbons sensing in atmosphere possibilities because the lidar equation (7) has been formulated for the laser single pulse regime. It is necessary to deal with the pulses sequence really in practice when the signal detection is led in the signal storing regime especially when lidar system operate at the limiting ranging distance. Such a result in the second part of work is exhibited in Figure 7 for the ethane molecule and the 1000 pulses signal adding and storing. It must be notice that such a regime does not give the appreciable increasing in ranging distance because the background increases too. The Raman signal over the background power exceeding is recorded for this case in Figure 7 as in the previous Figure 6 only at the ranging distance of 1.1 km from lidar.

Therefore the third part of this work is the Raman lidar potential possibilities estimation studies with the diode lasers for the paraffin molecules sensing in atmosphere for such molecules concentration level detection in atmosphere over industrial region [14].

The diode laser application in the spectroscopic experiments allows us to give the assumption about their using possibility in the laser systems for remote sensing. However such a lasers radiation peak power is about 100 W and we shall consider only variants with the signal storing at the laser radiation pulses high repetition rate [15]. The Raman scattering lidar optical layout for the studied hydrocarbon molecules is remained practically the same as is presented in Figure 5.

For the such a lidar potential possibilities estimation the lidar equations (1) computer simulation have been fulfilled for the hydrocarbon molecules vibration Raman back scattering with the concentration level of 10 ^{19} cm^{-3} in the ranging distance up to 1000 m. The diode lasers radiation the 405, 660, 810 and 940 nm wavelengths have been chosen for the sensing as earlier in [15].

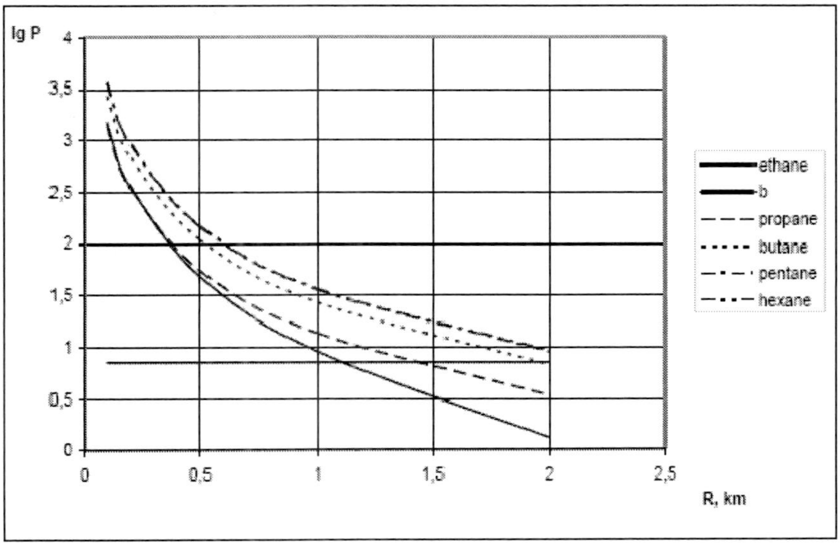

Figure 7. The plot of the studied molecules Raman power logarithm lgP (in nW) versus the ranging distance R dependence for 355 nm wavelengths laser radiation and molecules concentration of 1019 cm-3 for all molecules: ethane, propane, butane , pentane and hexane. The line b is parallel to the abscissa axis is the Sun background power level.

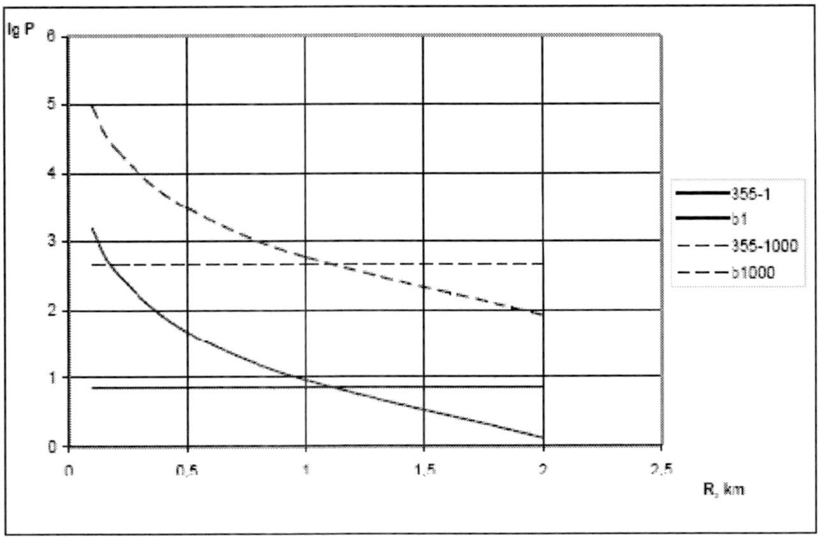

Figure 8. The plot of the ethane molecules Raman power logarithm lgP (in nW) versus the ranging distance R dependence for the 355 nm laser radiation wavelength and molecules concentration of 1019 cm-3 for two regimes: 1 pulse and 1000 pulses. The lines parallel to the absciss axe is the Sun light background power level for these two cases.

Table 4. The laser radiation and Raman bands wavelength, Raman differential cross section, PMT spectral sensitivity, atmospheric extinction coefficient and the Sun light spectral brightness values

λ_{0}, nm	λ_{R}, nm	$(d\sigma/d\Omega)\ 10^{30}$, cm^2/sr	$\xi_p(\lambda)$	k_0, km^{-1}	k_1, km^{-1}	$S(\lambda)$, $W/(m^2\ nm)$
methane (CH_4) =2914						
405	459,19	15,13	0,8	0,26	0,32	0,008
660	817,15	1,93	0,25	0,14	0,17	0,012
810	1060,25	0,64	0,13	0,13	0,09	0,005
940	1294,61	0,322	0,08	0,12	0,01	0,002
ethane (C_2H_6) =2954 $см^{-1}$						
405	460,03	22,53	0,8	0,26	0,31	0,008
660	819,83	3,19	0,23	0,14	0,17	0,012
810	1064,77	1,4	0,14	0,13	0,09	0,005
940	1301,35	0,77	0,09	0,12	0,02	0,002
propane (C_3H_8) =2886 $см^{-1}$						
405	458,61	38,45	0,81	0,26	0,31	0,008
660	815,29	4,91	0,26	0,14	0,17	0,012
810	1057,11	1,64	0,15	0,13	0,09	0,005
940	1289,84	0,82	0,9	0,12	0,02	0,002
butane (C_4H_{10}) =2890 $см^{-1}$						
405	458,68	43,94	0,81	0,26	0,31	0,008
660	815,55	5,61	0,26	0,14	0,17	0,012
810	1057,56	1,87	0,15	0,13	0,09	0,005
940	1290,6	0,935	0,9	0,12	0,02	0,002
pentane (C_5H_{12}) =2885 $см^{-1}$						
405	458,58	58,28	0,81	0,26	0,31	0,008
660	815,22	7,44	0,26	0,14	0,17	0,012
810	1057	2,48	0,15	0,13	0,09	0,005
940	1289,77	1,24	0,9	0,12	0,02	0,002
hexane (C_6H_{14}) =2886 $см^{-1}$						
405	458,61	62,98	0,81	0,26	0,31	0,008
660	815,29	8,04	0,26	0,14	0,17	0,012
810	1057,11	2,68	0,15	0,13	0,09	0,005
940	1289,84	1,34	0,9	0,12	0,02	0,002

The studied molecules Raman band back scattering differential cross sections have been calculated, as in [5], by the experimental cross section values for the 337 nm laser radiation wavelength and the chosen Raman bands from [8] and are also presented in Table 4. The PMT photocathode spectral sensitivity values $\xi(\lambda)$ for the studied molecules Raman bands wavelengths are also presented in Table 4. The atmospheric transmission along the ranging distance values have been calculated as earlier in [5] by the atmospheric extinction coefficient values k from [3, 5] and are collected in Table 4. The other parameters values in the lidar equation (7) were the same as in the previous case with the solid state laser.

The studied molecules Raman power numerical calculations have been fulfilled with these data at the concentration level of 10^{19} cm^{-3} in the chosen ranging distances from 10 up to 1000 m for all laser radiation wavelengths with peak power of 100 W and the laser pulses repetition rate values 10, 100 kHz and 1 MHz. The plot of the ethane molecules Raman power logarithm versus the ranging distance R as an example for all the laser radiation wavelengths is resulted in Figure 8. The background power has been calculated as in [5] with the Sun spectral brightness data from Table 4.

The Raman power values for other studied molecules and our experimental situation are even more due to the great values of these molecules Raman differential cross section as it follows from Table 4. The best result is got for the laser radiation 405 nm wavelength as it can be seen from Figure8. Therefore the best results for all the molecules and this 405 nm wavelength are collected in Figure 9.

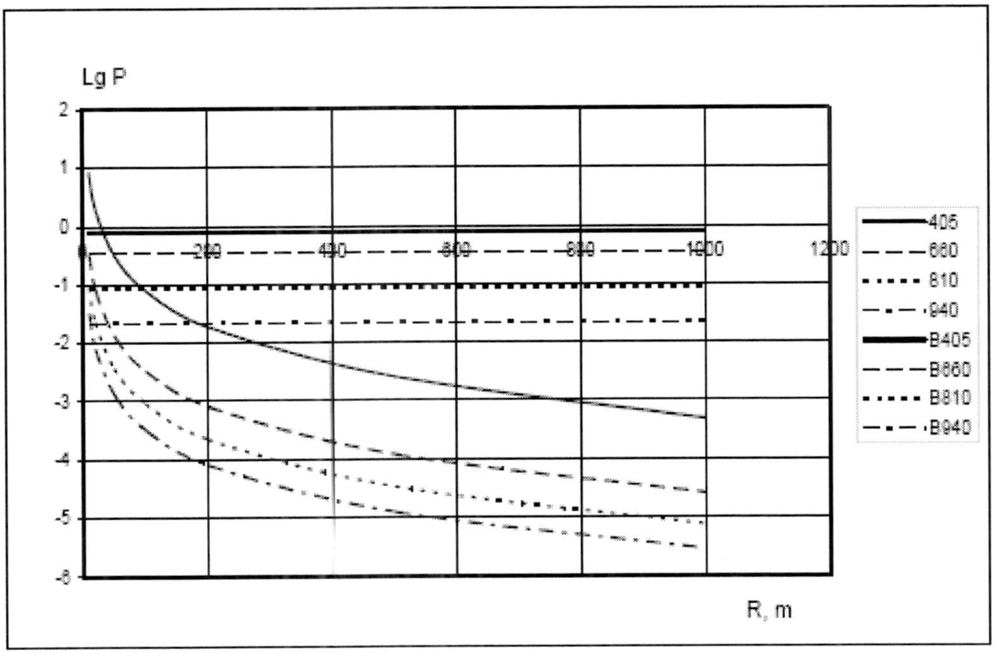

Figure 9. The plot of the ethane molecules Raman power values logarithm *lgP* (in nW) versus ranging distance *R* for all laser radiation wavelengths and molecules concentration of f 10^{19} cm^{-3}. The lines parallel to the abscissa axis B is the Sun background power level.

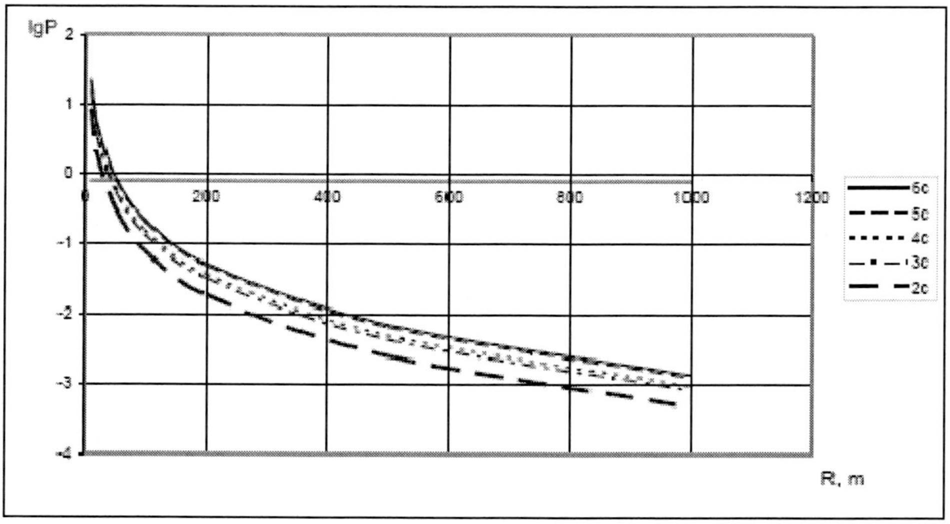

Figure 10. The plot of the studied molecules Raman power logarithm *lgP* (in nW) versus the ranging distance *R* dependence for 405 nm wavelengths laser radiation and molecules concentration of 10^{19} cm^{-3} for all molecules: ethane_-2C, propane – 3C, butane – 4C, pentane – 5C and hexane – 6C. The line B parallel to the abscissa axis is the Sun background power level.

The ranging distance increasing due to the diode laser peak power increasing is not obviously yet possible. Really it often deals with the diode laser generation in the pulse sequence more then 10 kHz. The detection is usually led in the synchronous photon counting regime in the fourth case and then the lidar systems can operate at the maximal ranging distances. And equation (7) can be used for the measurement time duration estimation needed for the studied molecules concentration level recording at the determined ranging distance. It is necessary to add new co-multiplier in the equation (8) N_i – the pulses number which is determined by the detection time *t*.

$$N_i = t f \tag{8}$$

where *f* – the pulses repetition rate.

Then the equation (7) will be seen as follows

$$E(\lambda, R) = E_0(\lambda_L) \cdot N_i \cdot \hat{E}_1 \cdot \Delta R \cdot A_0 \cdot T_L(\lambda_L, R) \cdot T(\lambda, R) \cdot N \cdot \frac{d\sigma(\lambda_L)}{d\Omega} \cdot \xi_p(\lambda) / R^2 \tag{9}$$

where $E(\lambda, R)$ – the recorded photon number at the photo detector and λ wavelength from the distance *R*; $E_0(\lambda_L)$ – the photon number in the laser radiation pulse and λ_L - its wavelength.

The photon number $E_0(\lambda_L)$ in the lidar equation (3) is determined by the expression:

$$E_0(\lambda_L) = \frac{E_L}{h \cdot \nu} \text{ or } E_0(\lambda_L) = \frac{E_L \cdot \lambda_L}{h \cdot c} \tag{10}$$

where E_L –laser pulse energy; h – Planck's constant; ν - laser radiation frequency; c – light velocity in vacuum.

Hence

$$t = \frac{E(\lambda, R)R^2}{E_0(\lambda_L)K_1\Delta RA_0T_L(\lambda_L, R)T(\lambda, R)N\frac{d\sigma(\lambda_L)}{d\Omega}\xi_p(\lambda)\nu_i} \qquad (11)$$

The 405, 560, 660, 810 and 940 nm diode laser radiation wavelength with the laser pulses repetition rates of 10 кHz, 100 кHz and 1 MHz and laser pulse energy about 1 mJ were chosen for the computer simulation as above mentioned. All other parameters are the same as in the lidar equation (7).

Using the above mentioned parameters value the detection time equation (11) computer simulation has been executed for the chosen laser radiation wavelengths and normal hydrocarbons in atmosphere with concentration about N = 10^{16} cm^{-3} sensing in the distance range up to 1 km and 10 Raman photons recording.

The plot of the detection time logarithm *lgt* (time *t* values in μs) versus ranging distance *R* up to 1000 m dependence for all the laser radiation wavelengths and the ethane molecules at the concentration level of 10^{16} cm^{-3} and laser pulses repetition rate of 10 кHz is shown as example in Figure 10.

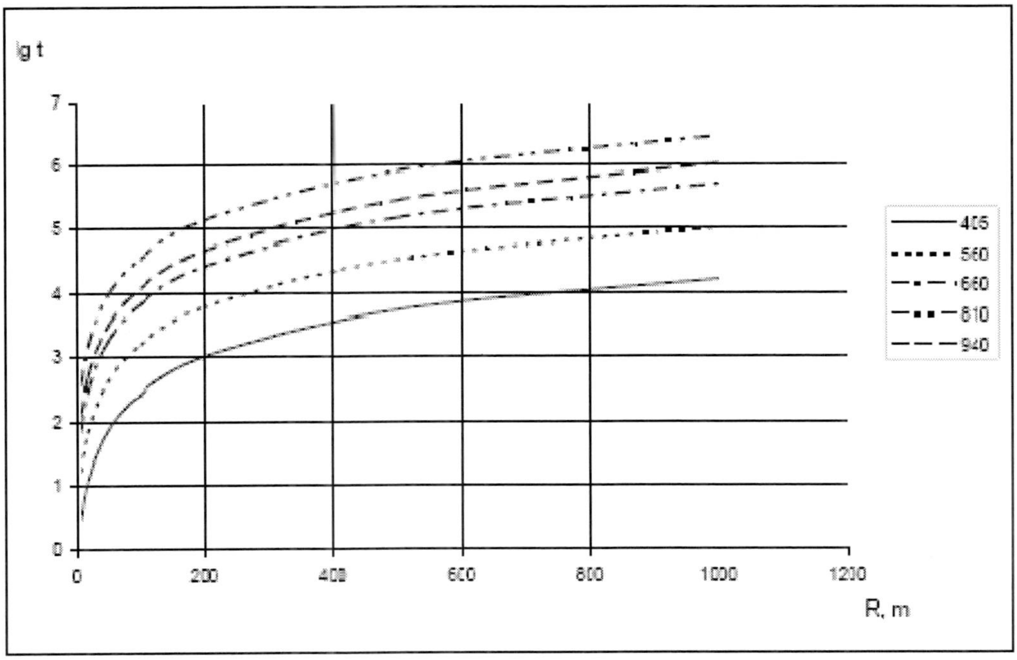

Figure 11. The plot of the detection time logarithm *lgt* (time *t* values in μs) versus ranging distance *R* for all the laser radiation wavelengths and the ethane molecules at the concentration level of 10^{16} cm^{-3} and laser pulses repetition rate of 10 кHz.

These results analysis shows that the optimal variant for the hydrocarbons sensing at the given concentration N_a and the Raman energy is equal to the 10 photons at the ranging distances up to 1 km is the laser radiation wavelength of λ_L = 405 nm (Figure 10). This variant allows to use the minimum detection time at the laser pulse energy of 1 mJ. So, it takes the detection time at the distance R = 1,0 km about 420 ms, at R = 50 m – 180 ms for the ethane molecules sensing with the concentration level of $N_a = 10^{16}$ cm^{-3} and 405 nm laser radiation wavelength. It was resulted at the 940 nm wavelength that it demands the detection time about 610 ms for ranging distance R = 1,0 km and 310 ms – for R = 50 m for comparison. And the detection time increases about 4 power with all the ranging distance longing from 0,01 to 1 km. If the FEU-79 PMT type photo detector counts more than 10 photons at the previously established detection time from ranging distance R=1,0 km and the determined laser pulses number generated into the atmosphere it can be told about the studied molecules concentration level excess the given level at this distance. Besides, these detection time values were satisfied to the lidar rapid recording requirement.

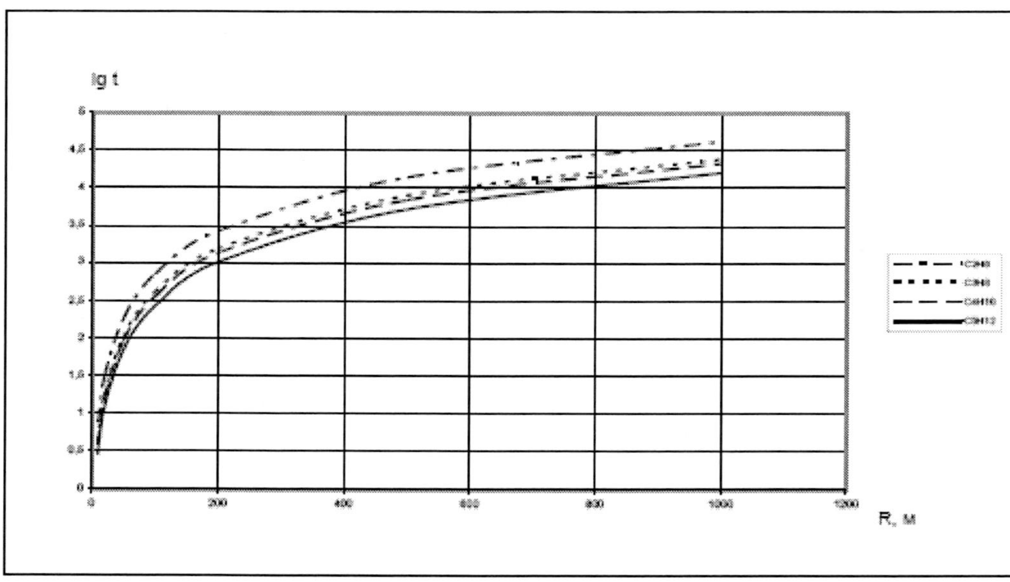

Figure 12. The plot of the detection time logarithm *lgt* (time *t* values in μs) versus ranging distance *R* for the 405 nm laser radiation wavelength and the studied molecules at the concentration level of 10^{16} cm^{-3}.

The detection time (in μs) logarithm *lgt* versus the ranging distance *R* dependence plots in the distance range up to 1000 m for the 405 nm wavelength and ethane, propane, butane and pentane molecules concentration level of 10^{16} cm^{-3} and laser pulses repetition rate of 10 kHz are exhibited in Figure 11. And the next detection time values have been calculated for the ethane molecules – 420 ms, the propane – 240 ms, the butane – 210 ms and the pentane – 180 ms for the ranging distance of 1 km that are quite correspond to the previous section results.

The further detection time reduction is possible at the laser pulses repetition rate increasing. It confirms by the plots in Figure 12. The detection time values have been decreased in the 6 powers in this case for the pentane molecules and the ranging distance of 1 km – from 180 ms at the repetition rate of 10 kHz to 18 μs at the repetition rate of 1 MHz.

As a whole, the Raman lidar equation computer simulation results show the possibility of the optimal laser radiation wavelength choosing for the paraffin molecules sensing in atmosphere in the photon- counting regime and these molecules concentration level exceeding over the given level in the distance range from 10 m up to 1 km.

Therefore, these results confirm the ability of the studied molecules concentration lidar measurements in atmosphere in the various experimental conditions. The 405 nm diode laser radiation was the optimal wavelength with the minimal detection time value of 180 ms for the pentane molecule at the ranging distance of 1km in the photon- counting Raman lidar regime.

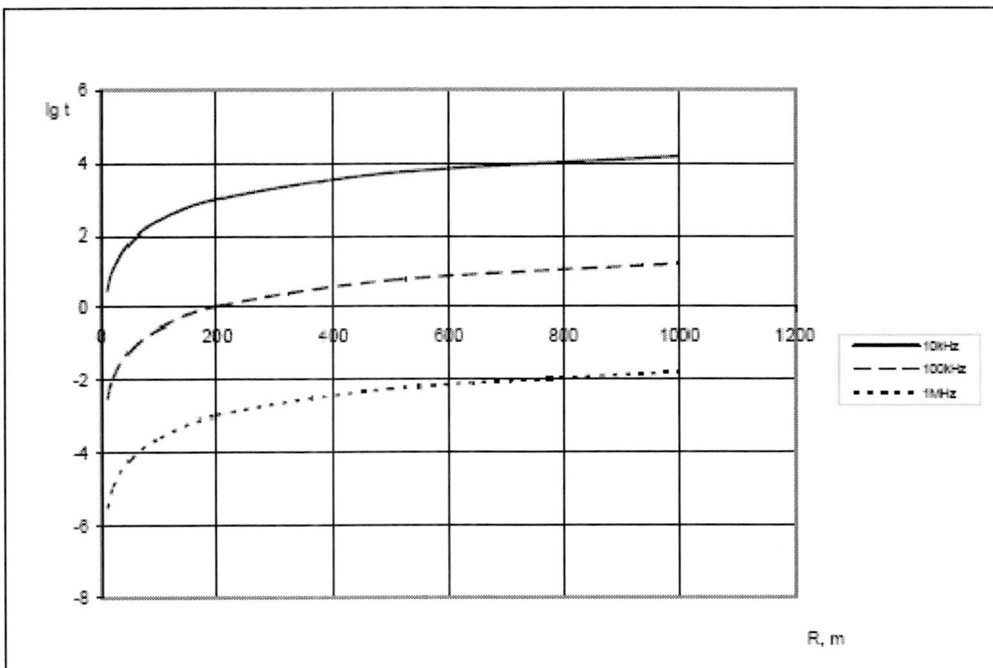

Figure 13. The plot of the detection time logarithm *lgt* (time *t* values in μs) versus ranging distance *R* for the 405 nm laser radiation wavelength and the pentane molecules at the concentration level of 10^{16} cm^{-3} and three laser pulses repetition rate values: 10, 100 kHz and 1 MHz.

4. The Hydrocarbon Molecules Concentration Remote Measurement by Space Raman Lidar

The pollutants emergency emission in the atmosphere sensing by the space platform lidar is the new perspective in the global environmental monitoring. Besides them this problem has been discussed in [3] only space shuttle aerosol lidar has been realized up to now [5]. The differential absorption and fluorescence lidars are examined as the possible lidar variants too [16]. Then Raman lidar potential possibilities theoretical estimation for the paraffin molecules emission in the near-Earth atmospheric layer at the concentration level of LPL that are equal to 10^{15} cm^{-3} and above detection from the space orbit presents the essential interest in this part of studies. This LPL for the paraffin molecule is equal about 10^{15} cm^{-3} [1]. As it has been shown earlier in [14, 16] the photon-counting photodetector devices have the maximal efficiency for the molecular Raman signals detection with such a concentration levels at the great distances from the lidar recording system.

Thus this part of paper goal is the photon-counting Raman back scattering lidar equation computer simulation for the CH_4, C_3H_8 and C_5H_{12} molecules vertical sensing in the near-Earth atmosphere [16] at the height range from 600 to 100 km from space orbit for the paraffin molecules concentration level sensing estimation.

The photon-counting Raman back scattering lidar equation was considered in the view (7) where the ranging distance R was changed by height H as in [12]. The YAG-Nd laser second, third and forth harmonics radiation wavelengths of 532, 355 and 266 nm with laser pulses repetition rate of 1 MHz and 10 ns time duration, 1 MW laser radiation pulse peak power and the lidar receiving telescope effective cross section of 0.785 m^2 were chosen for the simulation. The studied molecules Raman differential cross section values at the laser radiation wavelength $(\frac{d\sigma}{d\Omega})$ have been calculated with the known experimental data for the methane CH stretching: symmetrical at 2914 cm^{-1} and anti symmetrical – at 3017 cm^{-1} [3] the propane CH stretching: symmetrical at 2914 cm^{-1} and anti symmetrical – at 3017 cm^{-1} [3] and the propane CH stretching: symmetrical at 2914 cm^{-1} and anti symmetrical – at 3017 cm^{-1} [3] as in [17]. The results for all of the studied molecules CH symmetrical stretching were given in Table 5. The PMT photocathodes spectral sensitivity values $\xi(\lambda)$ for the studied molecules Raman bands calculated wavelengths were taken from [9] and were gathered in Table 5.

Table 5. The wavelength, Raman back scattering differential cross section, extinction coefficient and PMT photocatode spectral sensitivity values for the studied hydrocarbons molecules values

Molecule	**Methane CH_4**				**Laser**
ν, cm^{-1}	2914				
λ_0 , nm	$\lambda_{кр}$, nm	$(d\sigma/d\Omega)\cdot 10^{30}$, cm^2/sr	k, km^{-1}	ξ_p (λ)	k, km^{-1}
532	629,6	2,8	0,15	0,48	0,16
355	396,0	14,0	0,24	0,88	0,31
266	288,4	44,5	0,54	0,36	0,785
Molecule	**Propane C_3H_8**				
ν, cm^{-1}	2914				
532	629,6	13,2	0,15	0,48	
355	396,0	66,3	0,24	0,88	
266	288,4	210,0	0,54	0,36	
Molecule	**Pentane C_5H_{12}**				
ν, cm^{-1}	2914				
532	629,6	19,9	0,15	0,48	
355	396,0	101,0	0,24	0,88	
266	288,4	319,0	0,54	0,36	

The atmospheric transmittance were calculated as earlier in [5] by formula

$$T(\lambda, H) = \exp[-\int_0^R k(\lambda) dH] \quad (12)$$

with the extinction coefficient k from [9] and were taken in Table 5 too. The atmospheric extinction coefficient variation with height values H for the chosen laser radiation and Raman bands wavelength has been taken into account with the accordance of the Figure19.4 plot in [9, P.382].

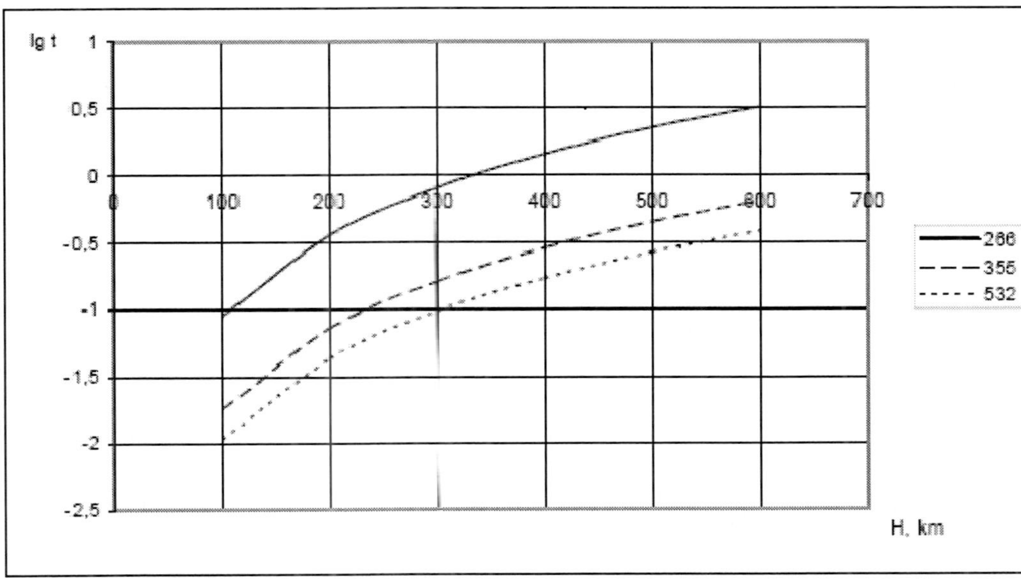

Figure 14. The plot of the recording time logarithm *lgt* in seconds versus height H in the range 100 ... 600 km dependence for the laser pulse repetition rate of 1 MHz, peak power of 1 MW, the third chosen laser radiation wavelength and methane concentration about 10^{15} cm^{-3}.

The lidar equation (10) computer simulation has been fulfilled with the above mentioned data for the chosen laser radiation wavelength and the studied methane molecules at the concentration level about $N = 10^{15}$ cm^{-3} in the distance range 100 ... 600 km for our experimental conditions and the possibility to record 1 Raman photon. These simulation results for this height range and all of the wavelengths were shown in Figure 13 and Figure 14. As it follows from Figure13 the minimal recording time was got for the 532 nm wavelength. The recording time was 0.39 s at this wavelength and sensing from 600 km height but it was increased up to 3.23 s at the 266 nm wavelength that is the maximal value for our experimental task. The photon- counting Raman back scattering lidar equation computer simulation results for the methane molecules sensing time duration from height 600 km to 100 km and laser pulse repetition rate of 1 MHz with peak power of 1 MW and methane concentration level of 10^{15} cm^{-3} can be exhibited in Table 6.

The laser peak power increasing up to 10 MW in the plot of Figure 14 for all of the wavelengths leads to the recording time decreasing in a power. The minimal value is 1.1 ms from 10.7 ms for the 532 nm wavelength and 100 km height but 38.5 ms from 385.2 ms for the 600 km height.

Table 6. The recording time *t* (in s) needed for the 1 photon recording

λ, nm	266	355	532
H, km	Time duration *t*, s		
100	0,09	0,018	0,011
300	0,81	0,16	0,096
600	3,23	0,65	0,39

The sum recording time along the 600 km all height is equal to 4 ms. We assume that laser radiation pulse goes forward and back again. This means that 808 laser pulses will be generated in this recording time of 3.23 s. The real recording time will be yet smaller and it is 300 μs when the one measuring time is equal to 1 μs in the suggestion that the studied molecules were really situated in the near-Earth atmospheric layer about 3 km.

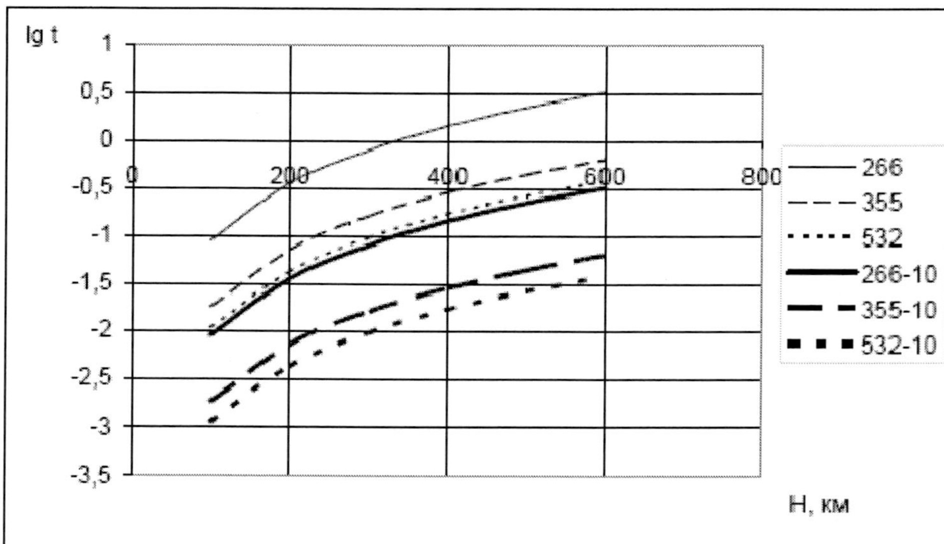

Figure 15. The plot of the recording time logarithm *lgt* in seconds versus height *H* in the range 100 … 600 km dependence for the laser pulses repetition rate values of 1 MHz, the methane concentration level about 10^{15} cm^{-3} , the third chosen wavelengths and laser pulse peak power two values of 1 and 100 MW.

The same lidar equation computer simulation has been fulfilled for propane molecules with the above mentioned data for our experimental conditions. These molecules have been studied at the concentration level about $N = 10^{15}$ cm^{-3} in the distance range 100 … 600 km and the lidar photodetector possibility to record 1 Raman photon. These simulation results for the height range and all of the chosen wavelengths were shown in Figure 15. As it follows from Figure 15 the minimal recording time was got for the 532 nm wavelength too. The sensing recording time was 0.082 s at this wavelength and sensing from 600 km height but it was increased up to 0.68 s at the 266 nm wavelength.

Furthermore, the same lidar equation (10) computer simulation has been fulfilled for the pentane molecules with the above mentioned data for our experimental conditions. These molecules have been studied at the concentration level about $N = 10^{15}$ cm^{-3} in the height range 100 … 600 km and the lidar photodetector possibility to record 1 Raman photon. These simulation results for this height range and all of the chosen wavelengths were shown in Figure 16. As it follows from Figure 16 the minimal recording time was got for the 532 nm wavelength. The recording time was 0.054 s at this wavelength and sensing from 600 km height but it was increased up to 0.45 s at the 266 nm wavelength. The Raman lidar equation (10) computer simulation results for the pentane molecules sensing time duration from these height values and laser pulse repetition rate of 1 MHz with peak power of 1 MW and pentane concentration level of 10^{15} cm^{-3} were exhibited in Table 7.

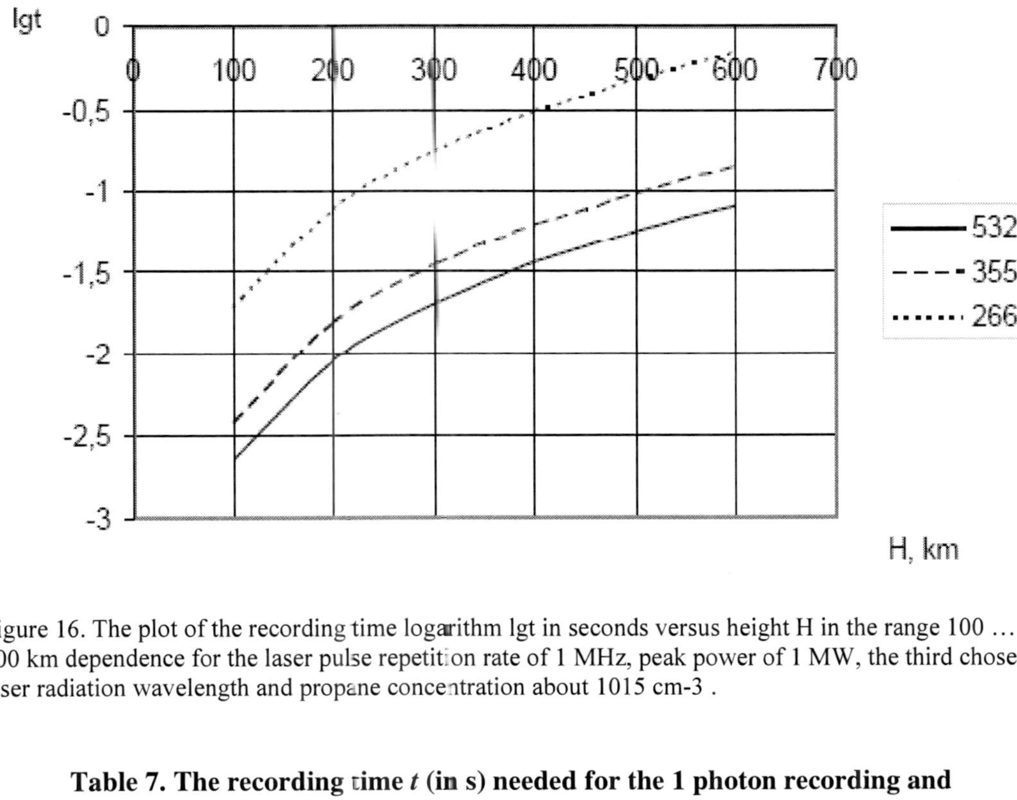

Figure 16. The plot of the recording time logarithm lgt in seconds versus height H in the range 100 … 600 km dependence for the laser pulse repetition rate of 1 MHz, peak power of 1 MW, the third chosen laser radiation wavelength and propane concentration about 1015 cm-3 .

Table 7. The recording time *t* (in s) needed for the 1 photon recording and pentane molecule

λ, nm	**266**	**355**	**532**
***H*, km**	**Time duration *t*, s**		
100	0,0015	0,0025	0,012
300	0,014	0,022	0,11
600	0,054	0,089	0,45

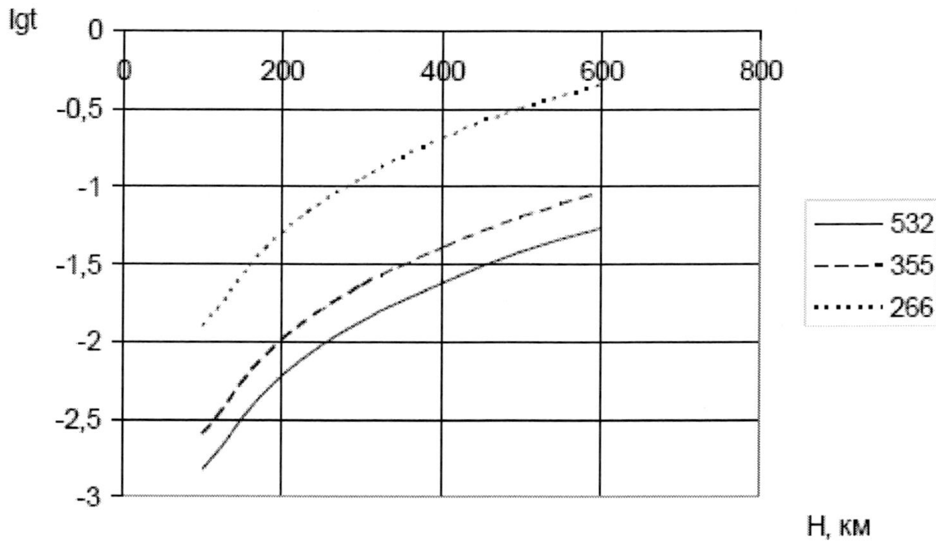

Figure 17. The plot of the recording time logarithm *lgt* in seconds versus height *H* in the range 100 … 600 km dependence for the laser pulse repetition rate of 1 MHz, peak power of 1 MW, the chosen laser radiation wavelength and pentane concentration about 10^{15} cm^{-3} .

These simulation results for these height values of 100 and 600 km and our experimental situation as the recording time logarithm versus the wavelength dependence plot were shown in Figure 17.

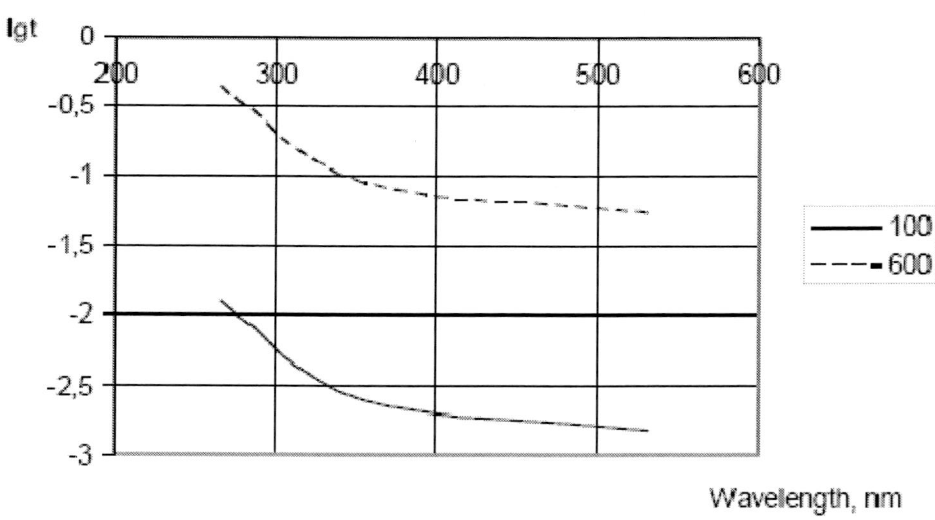

Figure 18. The plot of the recording time logarithm *lgt* in seconds versus laser radiation wavelength dependence for the laser pulse repetition rate of 1 MHz, peak power of 1 MW, the pentane concentration about 10^{15} cm^{-3} and .height H values of 100 and 600 km.

The Raman lidar equation computer simulation results for all of the studied molecules methane, propane and pentane molecules sensing time duration logarithm from these height values and laser pulse repetition rate of 1 MHz with peak power of 1 MW and these molecules concentration level of 10^{15} cm^{-3} were gathered in Figure 18. We can see from the plots in Figure 18 that the minimal studied molecules Raman differential cross section values at the laser radiation wavelength of 532 nm as the best result leads to the maximal recording time duration as for the methane molecules as an example.

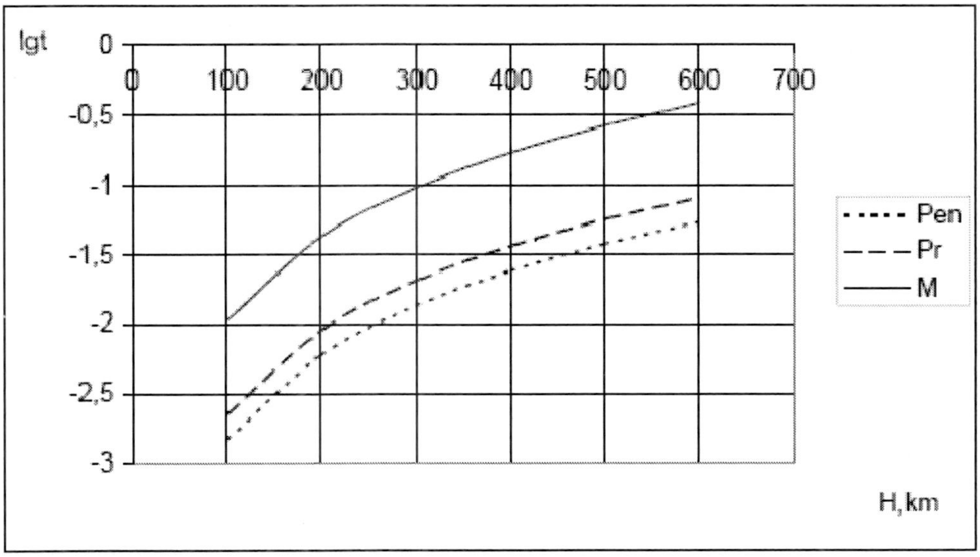

Figure 19. The plot of the recording time logarithm *lgt* in seconds versus height *H* in the range 100 … 600 km dependence for the laser pulse repetition rate of 1 MHz, peak power of 1 MW, the laser radiation wavelength of 532 nm and all studied molecules concentration about 10^{15} cm^{-3} .

The determined limitations will be applied to above simulation results by the lidar space platform moving character. This platform moves in the circle orbit with the velocity about first space of 7.9 km/s. And point of view on the Earth surface displaces due to its orbital movement with the velocity of 464 m/s and the angle to the space platform orbit plane. Then this point of view will displace into the one measurement time duration along the velocities parallelogram diagonal to the maximal distance about 25.5 km and minimal one about 3.2 km for the 600 km orbit height. Here the receiving telescope field of view with the divergence about 10^{-3} will be at the Earth surface as a circle with diameter of 601 m. Therefore the measuring volume for the minimal recording time of 0.39 s at the 600 km height and 532 nm wavelength for methane molecules will be equal to 2.1 km^3 for the atmospheric layer thickness of 150 m. Besides them the platform own will make one circle in 1.54 hour and the point of view at the Saint-Petersburg city latitude in this time duration will displace on 1276 km distance that will allow to fulfill the recording at the Russia territory only 7 times along the platform movement trajectory. And the measuring volume for the pentane minimal recording time of 0.054 s at the 600 km height and 532 nm wavelength for the studied molecules will be equal only to 0.29 km^3 for the atmospheric layer thickness of 150 m.

These results analysis shows that the recording time for the laser radiation wavelength λ_0 = 532 нм wavelength and height range 100 … 600 km will change from 11 ms up to 385 ms for this optimal lidar variant and the studied molecules concentration level about 10^{15} cm^{-3} .

Therefore, the computer simulation results show the possibility estimation of the optimal laser radiation wavelength choosing for the studied molecules sensing in atmosphere from space orbit in the photon-counting Raman back scattering regime at the concentration level exceeding 10^{15} cm^{-3} in the ranging distances from 600 to 100 km.

The optimal lidar variant choosing for the studied molecules sensing in atmosphere from space orbit for the 532 nm laser radiation wavelength gives the measuring volume for the studied molecules will be equal only to 0.29 km^3 for the atmospheric layer thickness of 150 m and the minimal recording time of 0.054 s from the 600 km ranging height.

CONCLUSION

This paper results allows the possibility of the laser radiation parameters optimal choosing for the paraffin concentration sensing in the atmosphere by the differential absorption and scattering lidar and Raman lidar at the required distance and concentration in the night and daytime condition.

The Raman lidar equation computer simulation results show the possibility of the optimal laser radiation wavelength choosing for the paraffin molecules sensing in atmosphere in the photon- counting regime and these molecules concentration level exceeding over the given level in the distance range from 10 m up to 1 km.

Furthermore, the computer simulation results show the possibility estimation of the optimal laser radiation wavelength choosing for the studied molecules sensing in atmosphere from space orbit in the photon-counting Raman back scattering regime at the concentration level exceeding 10^{15} cm^{-3} in the ranging distances from 600 to 100 km. These results analysis shows that the recording time for the laser radiation wavelength λ_0 = 532 нм wavelength and height range 100 … 600 km will change from 11 ms up to 385 ms for this optimal lidar variant and the studied molecules concentration level about 10^{15} cm^{-3} . The optimal lidar variant choosing for the studied molecules sensing in atmosphere from space orbit for the 532 nm laser radiation wavelength gives the measuring volume for the studied molecules will be equal only to 0.29 km^3 for the atmospheric layer thickness of 150 m and the minimal recording time of 0.054 s from the 600 km ranging height.

The experimental studies and computer simulation results were stated that the differential absorption and scattering lidar and Raman lidar have the preference possibilities in this problem solving as have been seen from our previous papers.

REFERENCES

[1] The atmospheric air pollutants list and codes. Saint-Petersburg. Scientific research institute "Atmospheric air protection". 1998, 161 pp.

[2] Privalov, V.E.; Chartiy, P.V.; Shemanin, V.G. *Life Safety* (Rus). 2003, No.9, 26-29.

[3] Measures, R. Laser remote sensing, Moscow: Mir, 1987. 550 pp.

[4] Privalov, V.E.; Shemanin, V.G. *J.Tech. Phys*. (Rus). 1999, vol.69, 65 – 68.

[5] Privalov, V.E.; Shemanin, V.G., Parametres of the gaseous molecules and the aerosol in atmosphere lidar remote sensing. Saint-Petersburg. Baltic State Technical University «VOENMECH». 2001, 56 pp.

[6] Ablyazov, E.K.; Shemanin, V.G. *Technospere Safety* (Rus). 2010, No. 4, 12-15

[7] Sverdlov, L.M.; Kovner, M.A.; Krajnov, E.P. Vibrational spectra of the multiatomic molecules. Moscow. Nauka. 1970, 560 pp.
[8] Voronina, E.I.; Privalov, V.E.; Shemanin, V.G. *Ecological systems and devices* (Rus.). 2002, No. 4, 13 – 15.
[9] Laser Handbook. Ed. Prokhorov, A.M. Moscow. *Sov. Radio*. 1978, vol. 1 and vol. 2.
[10] Inaba, H.; Kobaysi, T. *Opto - electronics*. 1972, vol. 4, 101 – 123.
[11] Voronina, E.I.; Privalov, V.E.; Shemanin, V.G. *Life safety*. (Rus.). 2003, No. 9, 30 – 33.
[12] Privalov, V.E.; Shemanin, V.G. *Proc. SPIE*. 2000, vol. 4064, 2-11.
[13] Laktjushkin, G.V.; Privalov, V.E.; Shemanin, V.G., *Tech. Phys. Letters*. (Rus.). 1998, vol. 24, 32 - 35.
[14] Ablyazov, E. K.; Shemanin, V. G. Optical Memory and Neural Networks (*Information Optics*). 2010, vol. 19, 201-205.
[15] Veremev, R.N.; Privalov, V.E.; Shemanin, V.G. *J. Tech. Phys*. (Rus.,). 2000, vol. 70, 115-118.
[16] Voronina, E.I.; Privalov, V.E.; Shemanin, V.G. Optical Memory and Neural Networks (*Information Optics*). 2009, vol. 18, 212–217.
[17] Privalov, V.E.; Shemanin, V.G. *Proc. SPIE*, 2001, vol. 4680, 106 – 111.

In: Air Pollution Emissions
Editors: Daniel G. Vasiliev and Robert A. Kask
ISBN: 978-1-62100-453-0

Chapter 3

THE POTENTIAL IMPACT OF MARINE AEROSOLS VIA THE SWELL AND THE OCEANIC WAVES ON THE PM10 CONCENTRATION MEASUREMENTS AT URBAN MARINE LOCATIONS

Chatrapatty Bhugwant[1], Miloud Bessafi[2] and Bruno Siéja[1]
[1] Observatoire Réunionnais de l'Ai -,
Technopole de La Réunion - Sainte-Clotilde - Réunion Island, France
[2] LE2P - Université de La Réunion -
Saint-Denis Messag Cedex 9 - Réunion Island, France

1. INTRODUCTION

Smoke and particles from vegetation and wildfires blazing around the world (America, Europe and Africa) have recently shown the potential environmental and health effects induced by these pollutants (*Liousse et al., 1995; Andreae and Merlet 2001; Crutzen and Andreae, 1990; Goldammer, 2003; Niemi et al., 2005*). The increasing interest in the study of atmospheric particles resides in their potential adverse effects on health, climate and the environment (*Dockery and Pope, 1994; Vianna et al., 2005*).

The biggest health threat comes from fine particles (Katsouyanni, 2003). Indeed, these microscopic particles can get into eyes and the respiratory system, where they cause severe health problems such as burning eyes, runny nose, and illnesses such as bronchitis (*Franklin et al., 2007; Pope et al., 2002, 2009*). Fine particles can also aggravate chronic heart and lung diseases and are even linked to premature deaths for the population exposed to these conditions (*Gehring et al., 2006*). The tiny particles are commonly associated with adverse health effects, and no threshold level has been recognized below which they have no impact on human health (*Laden et al., 2000; Pope et al., 2002; WHO, 2003*).

Atmospheric particles are emitted by both anthropogenic activities (ex. traffic circulation) and natural events such as volcanic eruptions and sea-spray (*De Leeuw, 1986; Bhugwant et al., 2000; Querol et al., 2009; Weijers et al., 2011*). They are either directly emitted into the air by diverse sources such as combustion process and windblown dust, or formed in the atmosphere by transformation of emitted gases such as SO_2 (sulphur dioxide) (*Seinfeld and Basset, 1982; Finlayson-Pitts and Pitts, 1999*). Concerning sea spray aerosol (SSA), it consists of particles in suspension in air that are directly produced at the sea surface, mainly in the liquid phase (i.e., as drops). The radii of these particles vary from ~10 nm to several mm (*De Leeuw, 2011*). The effects of exposure to any hazardous substance depend on the dose, the duration, the way one may be exposed, the personal traits and habits, and whether or not other constituents are present (*WHO, 2003*).

Particulate matter (PM) or particle pollution is a complex mixture of extremely small particles and liquid droplets. Particle pollution is made up of a number of components, including acids (such as nitrates, chlorates and sulfates), organic chemicals, metals, and soil or dust particles (*Penttinen et al., 2001; Querol et al., 2001*). These particles are found in different size ranges such as coarse, fine and ultra-fine. They also vary in composition and origin (*Okada, 1983; Friedlander, 2000; Osthoff et al. 2008*). The mass and the composition in urban environments tend to be divided into two principal groups: coarse particles and fine particles (*Niemi et al., 2005; Johansson et al., 2007*). The barrier between these two fractions of particles usually lies between 1 micrometer (μm) and 2.5 μm. However, the limit between coarse and fine particles is sometimes fixed by convention at 2.5 μm in aerodynamic diameter ($PM_{2.5}$) for measurement purposes. The smaller particles contain the secondarily formed aerosols (gas-to-particle conversion), combustion particles and recondensed organic and metal vapours (*Seinfeld and Bassett, 1982*).

The particles size is directly linked to their potential for causing health problems. Throughout the world, environmental and health organizations are much concerned about particles that are 10 micrometers in diameter or smaller because those are the particles that generally pass through the throat and nose and enter the lungs. Once inhaled, these particles can affect the heart and lungs and cause serious health effects (*EPA, 2003; ATSDR, 2008; Brunekreef et al., 2009*). In this sense, environmental and health organizations have grouped particle pollution into two categories, as follows:

"Inhalable coarse particles", such as those found near roadways and dusty industries, are larger than 2.5 μm and smaller than 10 μm in diameter.

"Fine particles", such as those found in smoke and haze, are 2.5 μm in diameter and smaller. These particles can be directly emitted from sources such as forest fires, or they can form when gases emitted from power plants, industries, automobiles and volcanic eruptions react in the air.

Inflammatory appearances of respiratory tracts, allergies, asthma attacks and chronic bronchitises, reduction of the vital capacity for the children, the increase of the cardiovascular mortality and the lung cancer for the adults, are some of the consequences identified for the pollution induced by airborne particles (*Don Porto Carero et al., 2001; Ruidavets, et al., 2005; Environment magazine, 2008*).

A recent study conducted in the European Union, in the frame of the CAFE (Clean Air For Europe) program showed that the fine and ultrafine particles contribute, every year, to the premature death of about 380 000 persons and to the loss of nine months of life expectation (*WICKS, 2002; EC-COM, 2005*).

The most evident effects on the environment and the economy directly linked to atmospheric particles are the spot on the buildings and the monuments, which cost billions of euros for their cleaning in mega cities each year (*Rabl et Teulère, 1999*). It is well known that particles decrease considerably the visibility (Ramanathan et al., 2002), as also recently observed in Russian capitals, during the July-August 2010 huge wildfires.

Hence, since a few decades, the surveillance of particles pollution is being reinforced in urban regions throughout the world, due to its effects on health and on the environment. Ambient particulate matter is responsible for harmful effects on health, even in the absence of other air pollutants (*Samet et al., 2000*).

Guidelines are recommended to be set for both short-term and long-term exposures to ambient particulate matter. However, current WHO and European Directives air quality guidelines describe the relationships between exposure to particulate matter and various health effects, but they recommend no specific maximum exposure values. Hence, new scientific evidence justifies reconsidering these relationships and developing guideline values both for fine and coarse particles and also revise the short-term duration exposure.

In this sense, the study of particulate pollution in urban regions located close to both anthropogenic (eg. traffic circulation) and natural (eg. marine, volcanoes ...) sources may help for a better comprehension of their impact on the air quality, the atmospheric chemistry, the environment and the human health. However, up to now, very few atmospheric measurements have been undertaken with links to sources such as urban and marine aerosols in the southern hemisphere, especially at Reunion Island (*Bhugwant et al., 2000; Bhugwant and Brémaud, 2001*).

The study of PM is thus of increasing interest to the scientific community (epidemilogists, researchers …) and to decision makers. Actually, one of the points subject to interrogation is the particles composition, due to a large variety of sources. This is due to the lack of atmospheric datasets in different environments and other parameters (eg. meteorological data, traffic density, etc.).

This chapter reports the PM10 concentration measured at LUT and BON, two urban air quality monitoring stations located in the Saint-Pierre city, to the South of Réunion Island. These two atmospheric stations are found close to each other (distant by 1.5 km) and to the sea (respectively at 500 and 200 m from it). The PM10 concentrations measured continuously with automatic analysers from August 2007 to August 2008 at these two sites were studied in conjunction with meteorological data (winds and rainfall) also recorded at Saint Pierre and the swell data recorded at Saint Pierre and other parts of the island.

2. Guideline Values for the PM10 Concentration Levels

As conferred previously, high PM10 concentration levels present in the ambient air may have considerable sanitary and environmental impacts (Pope et al., 2002; *EPA, 2003*). It is thus important to conduct the monitoring of this pollutant in densely inhabited regions, in order to take the adequate measures (eg. prevent the surrounding population, especially sensible persons such as children and old ones) if guideline values are exceeded and also to study its long-term trend and effects. In this sense, since a few decades, the WHO, the European community and the French Ministry of Environment (MEDDTL : Ministère de l'Ecologie, du Développement durable, du Transport et du Logement) have established guideline values for the PM10 concentration levels, following the severity of the impact of this pollutant on human health and on the environment.

The French national air quality criteria results mainly from the Decree n°2002-213, of the 15th February 2002 concerning the air quality survey, in particular for the PM10 and its effects on the human health and the environment, in the air quality objectives and in the limit values (*ADMINET, 2002*).

The European Council Directive (Council Directive, 1999) also defines the limit values for the PM10 in the ambient air. The objectives of this Directive are to establish limit values for particulate matter concentrations in ambient air intended to avoid, prevent or reduce harmful effects on human health and on the environment. The WHO guidelines imply not only guideline values, but also any kind of guidance given. Accordingly, for some substances the guidelines encompass recommendations of a more general nature that may help to reduce human exposure to harmful levels of air pollutants. For some pollutants, no guideline values are recommended, but risk estimates are indicated instead (*WHO, 2003*).

Table 1 resumes the main PM10 guideline values for the human health and for vegetation, published by the French Ministry of Environment, as the measurements presented and discussed in this study are undertaken at Réunion Island, a French county.

3. Experimental Set-Ups and Measurements

The PM10 concentrations have been recorded continuously at two close urban sites (distant by 1.5 km from each other) situated in the Saint Pierre city, since late 2007. This town is located to the South of Réunion Island, a tiny island located over the south-western Indian Ocean (see Figure 1). The PM10 measurements were subsequently compared with meteorological and swell data. The details concerning the datasets analysed in the frame of this study are described in Table 2. Figure 1 presents the map of Réunion Island, located over the south-western Indian Ocean in order to set the geographical context of the study.

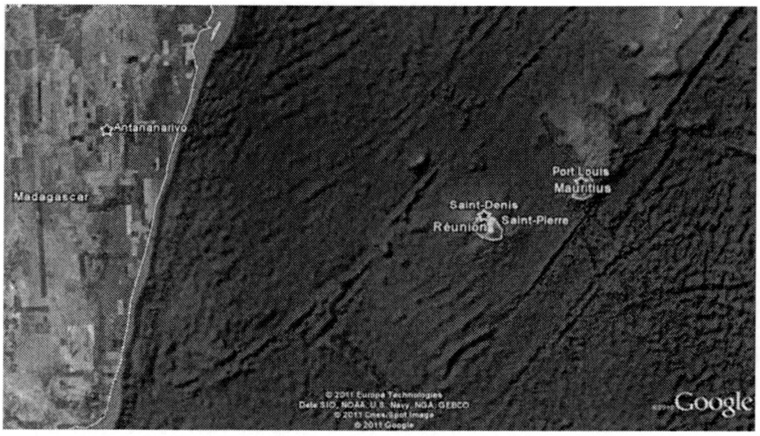

Source : GoogleEarth, 2011.

Figure 1. Map showing the geographic position of Réunion Island over the south-western Indian Ocean, with indication of Saint-Pierre city.

Table 1. Decree n° 2002-213 of the 15th February 2002 of the French Ministry of Environment relating the quality objectives and limit values for particulate matter, in ambient air

Pollutant	Type	Value	Period Considered	Calculation	Observations
PM10	Limit Value (LV) for the human health protection (exclusing natural events)	50 µg/m³	Civil year	Daily average	Centile 90.4: Not to exceed more than 35 days per year
		40 µg/m³	Civil year	Annual average	
	Quality Objective (QO)	30 µg/m³	Civil year	Annual average	

Quality Objective: A concentration level of polluting substances in the atmosphere, fixed on the basis of the scientific knowledge, with the aim of avoiding, preventing or reducing the harmful effects of these substances for the human health for the environment, to achieve for the given period.

Limit Values: A maximum concentration level of atmospheric pollutants, fixed on the basis of scientific knowledge, with the aim of avoiding, preventing or reducing the harmful effects of these substances for the human health and/or for the environment.

Limit values threshold for particulate matter - Adminet 2002

The limit values for the health protection used for the PM10 concentrations apply only to the part of the concentrations not linked to natural events. Are defined here by "natural events" the following ones: volcanic eruptions, seismic activities, geothermal activities, not cultivated land fires, violent winds, atmospheric put in suspension or transport of natural particles from desert regions.

These values are applicable since the 1st January 2005.

Table 2. Description (location, start date …) of the measurements undertaken at different regions over Réunion Island since 2007

N°	Station name	Abbrev. Name	Location/Region	Measurement	Measurement Dates
		ATMOSPHERIC MEASUREMENTS			
1	Bons Enfants	BON	Saint Pierre / South	Hourly NO_2 and PM10	August 2007 to August 2008
2	Luther King	LUT	Saint Pierre / South	Hourly NO_2 and PM10	August 2007 to August 2008
		METEOROLOGICAL MEASUREMENTS			
1	Pierrefond	PIE	Saint Louis / South	Hourly winds (direction & speed) and rainfall	January 2007 to August 2008
2	Grand Bois	GBO	Saint Pierre / South	Hourly rainfall	January 2007 to August 2008
3	Luther King	LUT	Saint Pierre / South	Hourly winds (direction & speed)	August 2007 to August 2008
		SWELL MEASUREMENTS			
1	Port Ouest	PO	Le Port / North-West	Hourly swell	August 2007 to August 2008
2	Port Ouest	PDG	North-West	Hourly swell	August 2007 to August 2008

Figure 2 presents a zoom on Réunion Island, with indication of the atmospheric, meteorological and swell measurement sites.

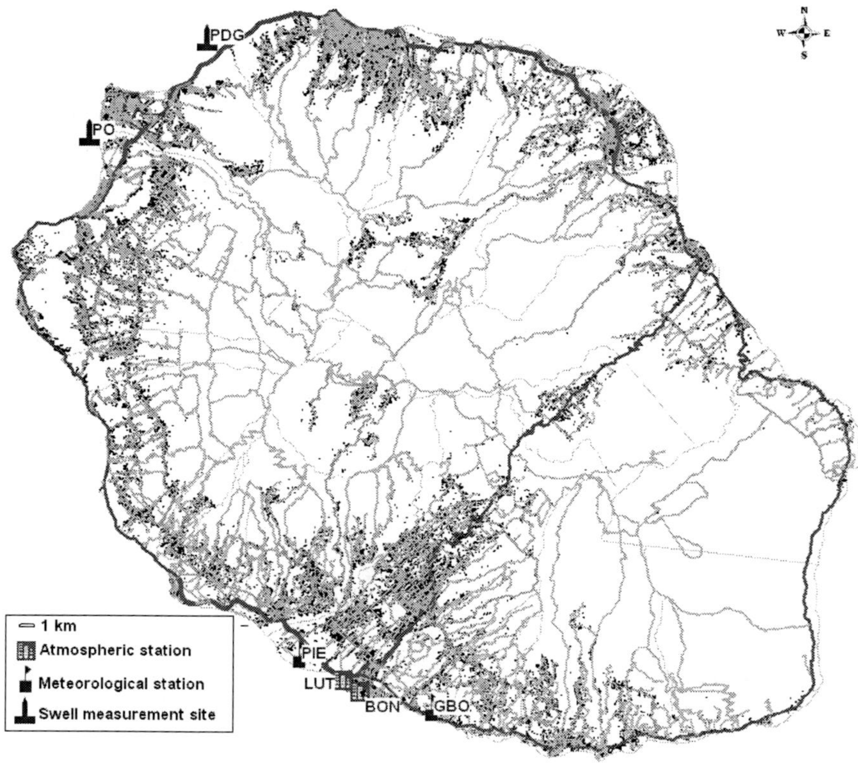

Source : IGN (Institut Géographique National), Paris.

Figure 2. Map of Réunion Island with indication of the atmospheric, meteorological and swell measurement sites dedicated to this study.

3.1. Atmospheric Data

The PM10 concentrations measured continuously at the LUT and BON atmospheric stations from August 2007 to August 2008 are analysed in this study. Indeed, since the implementation of these two measurements stations, they regularly present different levels and significant variability, although both sites are representative of a similar environment. These two stations are found at Saint Pierre, a city holding a population of about 80 000 inhabitants and exempt of heavy industries emitting atmospheric pollution (*INSEE, 2010*).

The two measurement stations are located in the same urban environment, with light traffic circulation and at more than 0,5 to 1 km away from heavy highways, and thus exempt of local anthropogenic contamination.

In order to establish the origin and main causes of the PM10 concentration variability recorded at both LUT and BON, the nitrogen dioxide (NO_2) concentration, an atmospheric tracer of anthropogenic origin (mainly traffic circulation, in absence of any other source such as heavy industries and/or volcanic eruption) (*Bhugwant et al., 2009*), was also monitored in parallel and analysed.

3.1.1. PM10 Measurements

The PM10 concentration has been measured at LUT and BON by a Rupprecht & Patashnick analyser, commercialised by Thermo Environmental Instruments, from August 2007 to August 2008. This instrument uses the TEOM (Tapered Element Oscillating Microbalance) technique for the PM10 analysers operating in the atmospheric monitoring network administered by ORA (Observatoire Réunionnais de l'Air) at Réunion Island. In order to measure also the volatile parts of the particles, the classical PM10 analyser was equipped with an FDMS (Filter Dynamics Measurement System) case in order to capture the volatile particles as well. The FDMS is a self referencing airborne particulate monitor based on the TEOM technology (*O'Neill et al., 2002; Wanjura et al., 2008*). The TEOM uses first principle physics to measure the mass of particles collected on a filter. The FDMS measures the core and volatile fractions of the collected mass by using a self referencing technique to measure the effects of these particles as they collect on the filter. Core particulate matter has an additive effect whereas semi-volatile particles can both add mass as they arrive on the filter and can later lose mass as they volatilise. Gases and vapours (including water vapour) can also have both positive and negative effects on the mass of the filter and thus the collected PM over time. The FDMS is able to monitor and report these changes (called the filter dynamics) as they occur and provide a more accurate and 'true' measurement of airborne particulate mass concentration.

The TEOM/FDMS measurement procedure allowed to follow-up continuously the PM10 concentrations at a 1 hour time-base since August 2007 at the Saint Pierre measurement locations.

3.1.2. Determination of the Aerosols Composition

The particles collected on the TEOM/FDMS filters at LUT and BON have subsequently been analysed in laboratory (at Micropolluants Technologie laboratory) using the on-line ion chromatography coupled with ICP-AES (Inductively Coupled with Plasma–Atomic Emission Spectrometry) method. This analysis allowed the determination of the concentration of aimed constituents, i.e., the chloride (Cl^-) ions, which is the main tracer of marine aerosols via the sea spray.

3.1.3. NO_2 Measurements

The NO_2 concentration was measured by a 42C Chemiluminescence NO-NO_2-NOx analyser model, commercialized by Thermo Fisher Scientific, at both LUT and BON from August 2007 to August 2008, in order to be compared with the PM10 concentration variability.

This Instrument uses the chemiluminescent photometric principle for the measurement of nitrous oxide (direct reading) and nitrogen dioxide (converted to NO for detection) emissions. This NOx monitor generates three continuous signals: NO, NO_2 and NOx (*Bhugwant et al., 2001*). The instrument analyses ambient air at a time-base of 15 minutes, subsequently averaged to 1 hour, for analysis with the other datasets, i.e., PM10, swell, winds and rainfall.

3.2. Meteorological Data

The local meteorological parameters such as wind speed, wind direction, and rainfall are measured from the instruments onboard a tower at 10 m above the ground level in meteorological stations located close to the LUT and BON atmospheric locations. These instruments are part of the French Meteorological Service (*Météo-France, 2000*). The meteorological parameters are measured since a few decades at several stations over the island. However, only data collected since August 2007 at Pierrefond (PIE) and Grand Bois (GBO) meteorological stations, which are located in the vicinity of the Saint Pierre agglomeration, are analysed in the present study.

The meteorological data measured continuously with a 1/2 hour time-base were subsequently brought to a 1-hr basis in order to be compared with the atmospheric (PM10 and NO_2 concentrations) and swell data.

3.3. Swell Data

As for the oceanic waves, the swell can be measured from the surface rise, with swell-graph buoys, swell perch and swell radars, due to the under water pressure (high frequency tide-graphs) or the water speeds (water current meters).

In this study, the swell data was measured by the Direction Départementale de l'Equipement (DDE) with a swell-graph fixed with a buoy in the sea, at a few meters away from the beach. The swell data are measured at several sites located on the Northern, North-Western and Southern parts of the island since 2006.

However, only data recorded from August 2007 to August 2008 at Port Ouest (PO) and Pointe du Gouffre (PDG) are considered in the present study (**Figure 2**). The swell data measured continuously at distinct locations close to Réunion Island with a 1 hour time-base, were analysed in conjunction with the atmospheric and meteorological data.

4. Results and Discussion

The analysis of the atmospheric and meteorological datasets is first focused on the diurnal variation of the diverse parameters, in order to study their trend and levels and also to look for possible links between them. The datasets are subsequently analyzed on a daily basis, in order to look closely at their variability and also to compare with the PM10 guideline values.

4.1. Diurnal Variation during August 2007-August 2008

4.1.1. PM10 Diurnal Concentration Variation

Figure 3 shows the mean diurnal variation of the PM10 concentration recorded from August 2007 to August 2008 at the LUT and BON air quality monitoring stations. It may be noticed that the PM10 diurnal trend and pattern is similar at both sites, with low values recorded at night-time (20 and 26 μg/m^3 respectively at LUT and BON) and high values measured at day-time (30 and 55 μg/m^3 respectively at LUT and BON). The PM10 concentration levels at BON are however systematically and significantly higher than at LUT. The difference between the PM10 concentrations recorded at BON is higher than at LUT by ~ 6 μg/m^3 at night-time (8:00 p.m. to 6:00 a.m.) and by ~ 25 μg/m^3 at day-time (7:00 a.m. to 7:00 p.m.). The PM10 enhancement is mainly observed at day-time on both sites, during which local anthropogenic activities are highest, while at night-time (when anthropogenic activities are negligible), the difference between the PM10 concentration at BON and LUT is lower.

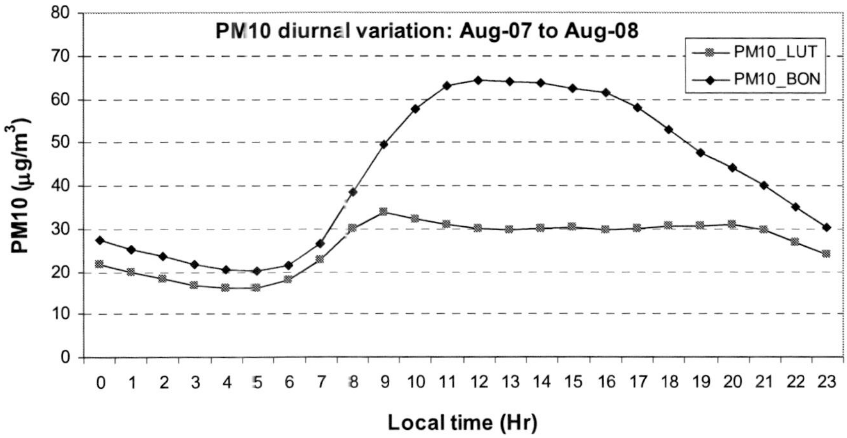

Figure 3. Mean diurnal variation of the PM10 concentration calculated from August 2007 to August 2008 at LUT and BON atmospheric stations.

Nevertheless, as the two locations are representative of the urban pollution prevailing over the Saint Pierre city and close to each other, they are expected to record the same levels and order of magnitude of PM10 concentrations at both sites, which is not the case. Importantly, there is not any industry emitting atmospheric pollutions at Saint Pierre especially in the vicinity of LUT and BON and both atmospheric stations are quite equally distant from light/heavy highways.

The monthly mean PM10 concentration profile (not shown) indicates that during austral winter (August to December 2007), the PM10 concentration is quite constant, with a mean value of 24 μg/m^3 and 42 μg/m^3 at LUT and BON respectively. During austral summer (January to May 2008) however, higher PM10 values are measured (average PM10 concentrations: 28 μg/m^3 and 48 μg/m^3 at LUT and BON respectively), with a maximum peak observed in February 2008 at both sites.

Importantly, the quality objective (QO = 30 μg/m^3, on annual average) as well as the limit value for the protection of the human health (Centile 90,4 = 50 μg/m^3, on daily average not to be exceeded more than 35 days/year and LV = 40 μg/m^3, on annual average) are exceeded at BON, when the datasets are analysed on a civil year.

We therefore tried to look for the origin and main causes of the significant difference between the PM10 concentrations recorded at LUT and BON.

In this sense, the meteorological parameters (winds and rainfall) were analysed in order to determine the sources/sinks of the atmospheric particles. Indeed, these parameters may also influence the atmospheric concentration variability, via dispersion/deposition and scavenging processes (*Byrne and Jennings, 1993; Uematsu et al., 2000*). We also analysed other atmospheric pollutants such as NO_2, which is a tracer of anthropogenic emissions (mainly traffic circulation, in this case).

4.1.2. Winds Diurnal Variation

The winds (direction and speed) data recorded from August 2007 to August 2008 at the Pierrefond (PIE) meteorological station have been analysed, as this station is best representative of the meteorological conditions which prevail at LUT and BON, due to its closeness with Saint Pierre city. The Figure 4 presents the mean winds (direction and speed) diurnal variation recorded from August 2007 to August 2008 at PIE.

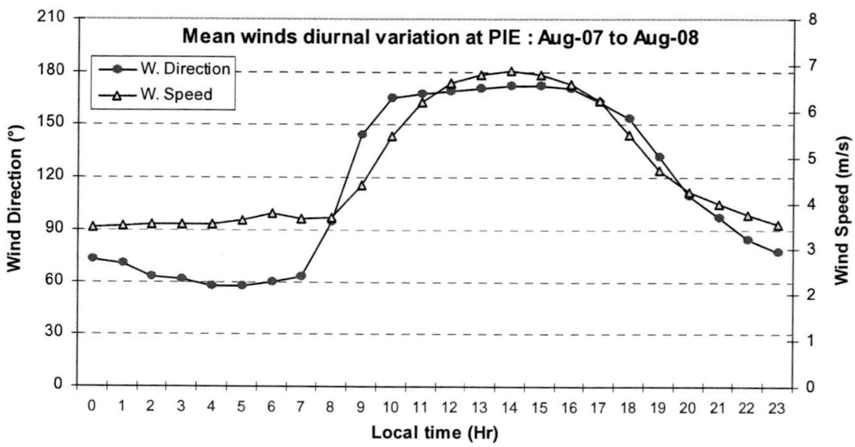

Figure 4. Mean winds diurnal variation calculated from August 2007 to August 2008 at Pierrefond (PIE) meteorological station.

The winds direction and speed exhibit quite the same trend and pattern at Saint Pierre. During day-time, the air samples have an eastern to southern origin (90-180°, i.e. mainly marine sector), with moderate-to-high speeds (~6 m/s) while at night-time, they have a north-eastern to eastern origin (60-80° sector, i.e., mainly inland/urban sector), with low speeds (~3,8 m/s). Due to the geographic configuration of the island and particularly of the LUT and BON locations, the regional atmospheric circulation (i.e. easterlies) is coupled with local circulations (land/sea breezes). The winds diurnal variation is mainly due to the land-sea interactions, in particular to the sea breezes at the surface, which contribute to local circulations (*Stull, 1988*; *Cheng, 2002)*.

The comparison of Figures 3 and **4** datasets point out that the night-time PM10 concentrations recorded at LUT and BON originate mainly from inland, while the day-time PM10 concentrations are mainly of marine origin, mostly at BON (~200 m away from the sea), as it is closest to the sea than LUT (~500 m from the sea)..

The results suggest that the day-to-night discrepancy in the diurnal PM10 concentration variation observed at LUT and BON may not only be due to local/regional anthropogenic activities but to a regular contribution of marine aerosols as well, especially at BON, which may explain the higher background PM10 levels observed at this measurement site.

4.1.3. NO_2 Diurnal Variation

The NO_2 concentration, an atmospheric tracer of anthropogenic activities (mainly traffic circulation at Saint-Pierre city) recorded at LUT and BON was also analysed in order to be compared with the PM10 concentration variability. The Figure 5 presents the mean NO_2 diurnal concentration variation recorded from August 2007 to August 2008 at LUT and BON.

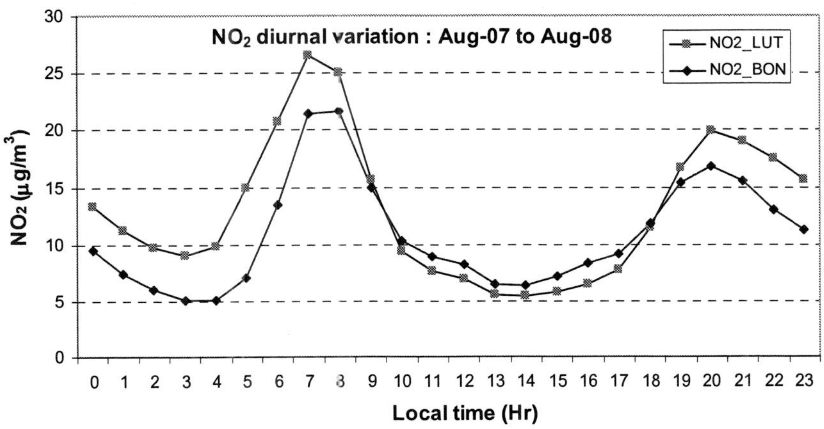

Figure 5. Mean NO_2 diurnal concentration recorded from August 2007 to August 2008 at LUT and BON atmospheric stations.

The mean diurnal NO_2 concentration variation exhibits the same trend and pattern, with the same order of magnitude at both LUT and BON. At night-time, when local anthropogenic activities are negligible, the mean NO_2 concentration is low-to-moderate (14 and 10 $\mu g/m^3$ at LUT and BON respectively).

During morning rush hours (7:00-9:00 a.m.) however, a maximum peak is observed (26 and 21 $\mu g/m^3$ at LUT and BON respectively). At day-time, when local anthropogenic activities are moderate, the mean NO_2 concentration is moderate-to-low (9 and 5 $\mu g/m^3$ at LUT and BON respectively). Then, during the evening rush hours (6:00-9:00 p.m.), a secondary peak is observed (20 and 17 $\mu g/m^3$ at LUT and BON respectively).

The NO_2 levels measured over Saint Pierre suggest that the traffic circulation, the main anthropogenic activity at this city affects only slightly the air quality. The anthropogenic source affects similarly LUT and BON, as the NO_2 concentration measured at both sites exhibit quite the same level and trend. Also, the notable difference observed between the NO_2 and the PM10 concentration variability at LUT and BON points out that various and distinct sources might influence the two measurement locations, although they are quite close to each other.

The results indicate that the day-to-night discrepancy in the diurnal NO_2 concentration variation observed at LUT and BON is low. Importantly, the local anthropogenic activities contribution on the atmospheric (NO_2) measurements is similar at both LUT and BON, thus confirming our previous assessments concerning the main origin (both inland/urban and marine) of the atmospheric particles.

4.1.4. Rainfall Diurnal Variation

The rainfall data recorded from August 2007 to August 2008 at Pierrefond (PIE) and Grand Bois (GBO) meteorological stations (see Figure 2) were also analysed. These two stations are located on both sides (west and south-east respectively) of Saint Pierre and are thus representative of the rainfall activity on local/regional-scale, i.e., over the Saint Pierre agglomeration as a whole, and particularly at LUT and BON. In general, the rainfall level is quite low over this region of the island. The rainfall level is variable (~0,1 mm) during day-time and moderate at night-time (~0,2 mm) at Saint Pierre (not shown).

The analysis of the monthly rainfall indicates a seasonal trend, with moderate but insignificant rainfall levels in austral summer (0,03-0,2 and 0,1-0,4 mm at PIE and GBO respectively) and low levels in austral winter (0,01-0,08 and 0,01-0,1 mm at PIE and GBO respectively). The results imply that this parameter might have some but minor influence on the PM10 concentration variability at Saint Pierre and particularly at BON. They also point out that the scavenging processes affect evenly the whole Saint Pierre agglomeration, and have insignificant effects on the PM10 measured at both LUT and BON locations.

4.1.5. Wind Roses of PM10 Diurnal Variation

Figure 6 presents the wind roses (from PIE winds data) of the PM10 diurnal variation measured at LUT and BON from August 2007 to August 2008.

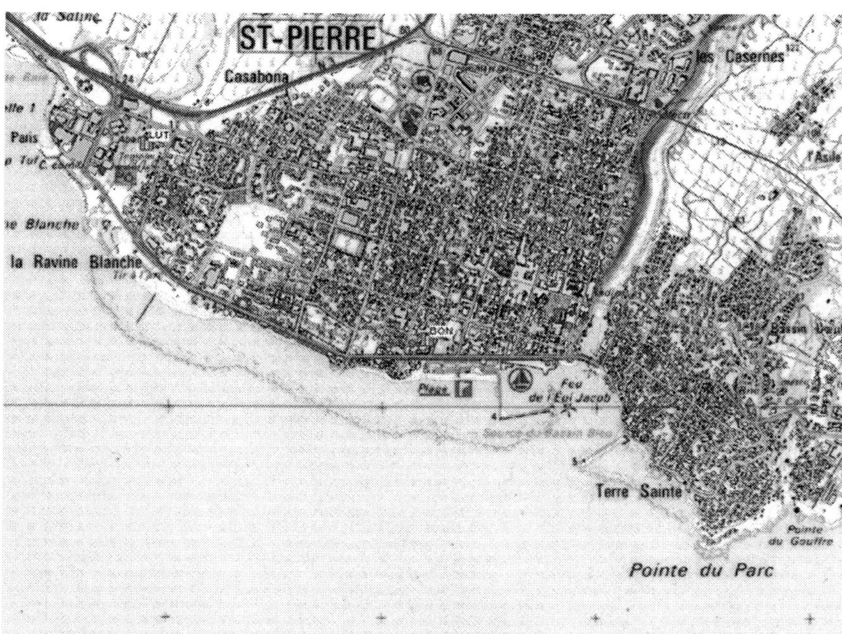

Figure 6. Wind roses of the PM10 diurnal variation calculated from winds data measured at LUT and BON from August 2007 to August 2008.

The wind roses diurnal variation of the PM10 measured at LUT and BON exhibit the same trend and pattern. The lowest PM10 concentration recorded at LUT and BON originate mainly from the north-eastern to eastern (60-110°) sectors (i.e. inland/urban region) at night-time, while the highest PM10 concentrations originate mainly from the eastern to southern (120-180°) sectors (i.e. marine region) at day-time.

However, since 7:00 a.m., the PM10 concentration recorded at BON exhibits a notable enhancement, as compared to that measured at LUT. Indeed, the mean PM10 concentration originating from the 90-180° sectors is 30 μg/m^3 at LUT, while it is 60 μg/m^3 at BON (Figure 3). It may therefore be assessed that marine aerosols are superimposed to the anthropogenic pollutants and affect notably the PM10 concentration recorded at BON. The results point out that although the PM10 measured at LUT may also in part be influenced by the marine aerosols (see Figure 6) it may as well be influenced by dispersion/deposition processes (eg. the vegetation, the surrounding buildings and the distance from the sea) than at BON.

4.1.6. Swell Diurnal Variation

The swell data was lacking at some measurement sites, during the study period. In particular, no swell data was available for the Saint Pierre region in 2008. As the swell is a regional and semi-annual phenomenon, data collected at other regions were analysed. However, in order to ensure that this extrapolation was valid for our study area, the monthly mean swell data recorded at different regions from August 2007 to August 2008 were analysed.

The monthly swell height shows a flat profile, with a mean value of ~1,2 m at both PO (not shown) and PDG from August 2007 to December 2007 (see Figure 8). A significant swell peak is observed on every site, with a comparable level at both sites (~1,8 m) in February 2008. The monthly mean swell is then of the same order of magnitude (~1,2 m) at all the sites, from March 2008 to August 2008, with a secondary peak (~1,3 m) in June 2008.

The results point out that the swell is not a localized phenomenon, and it does not attain the coasts of Réunion Island in a different way and does not concern specifically one sector of the island only. It is indeed a regional phenomenon touching globally and quite the same way all the coasts of the island. In this sense, PO corresponding to the site having most of the 2007-2008 swell data was analyzed.

Figure 7 presents the mean swell diurnal variation during August 2007-August 2008 at the Pointe du Goufre (PDG) site located to the North-West of Réunion Island.

The swell diurnal variation exhibits a similar profile as for the PM10 measured at BON, with a maximum level (mean : ~1,35 m) during day-time, when wind speeds are highest (due to the conjugate effect of sea breezes and easterlies) and a minimum at nighttime ((mean : ~1,24 m). The swell activity is a continuous phenomenon throughout the day and exhibits a seasonal variation throughout the year.

Also, due to the geographic configuration of the island (important relief) and the existence of coral reefs downwind the BON atmospheric station, the sea waves break constantly over the coral reefs and might thus continuously generate marine aerosols via sea spray. This process might explain the higher background PM10 levels systematically observed 'off' the intense swell activity (which occurred in February 2008) at BON than at LUT. For instance, from August 2007 to January 2008, the monthly mean swell was ~1,2 m high at both PO and PDG and correspondingly the monthly mean PM10 concentration was 24 and 42 μg/m^3 at LUT and BON respectively.

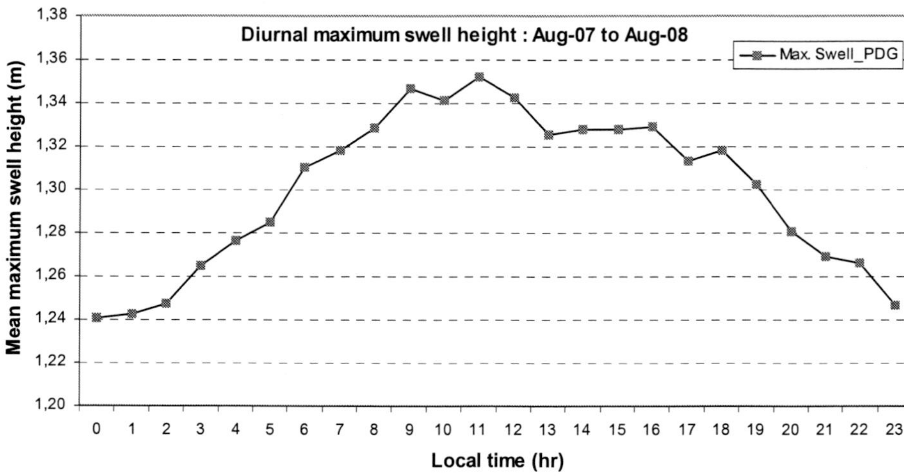

Figure 7. Mean swell diurnal variation recorded from August 2007 to August 2008 at Pointe du Gouffre (PDG) location.

The results point out that when the waves break over the coral reefs situated downwind the BON atmospheric station they generate marine aerosols which are transported preferentially and are superimposed on the background urban particles at BON (same order of magnitude as that recorded at LUT). During swell episodes and tides however, this process is reinforced, during which the PM10 levels measured at BON are much higher than at LUT. Thus, the conjunction of the winds and the swell/tides variation contribute to a mean enhancement of the PM10 level at BON by ~6 μg/m^3 at night-time and by ~30 μg/m^3 at day-time, as compared to the PM10 level measured at LUT (Figure 3).

4.1.7. Correlation between Swell and PM10 Variation

It was important to determine the influence of the swell activity on the PM10 concentration variability at BON. In this sense, a statistical analysis was performed in order to exclude events leading to the PM10 enhancement that might not be attributable to the swell activity. The Figure 8 shows the monthly mean swell height measured at PDG and the monthly mean PM10 concentration measured at BON from August 2007 to August 2008. It may be seen that there is a good correlation between the high swell activity and the high PM10 concentration in particular in January 2008 and February 2008. During September 2007 and August 2008 however, when the PM10 exhibit moderate levels, the swell activity is low, pointing out that other processes (anthropogenic sources, oceanic waves, wind speeds, etc.) may also in part influence the PM10 variability observed at BON. A close look at the LUT and BON NO_2 concentration profiles indicate that their pattern, trend and level is similar, suggesting that the anthropogenic contribution is identical at both sites. Hence, it may be inferred that the oceanic waves may also in part contribute to the background ('off swell events') PM10 enhancement at BON, via the sea spray, during November and December 2007.

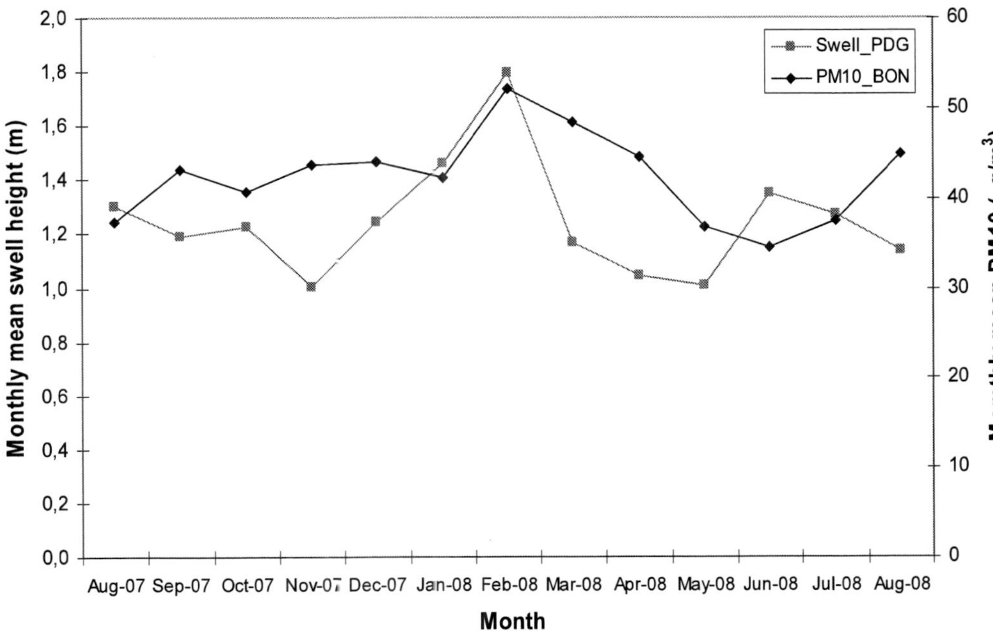

Figure 8. Monthly mean swell height measured at PDG and monthly mean PM10 concentration measured at BON from August 2007 to August 2008.

4.3. PM10 Concentration Variation during August 2007-August 2008: Comparison with Guideline Values

Figure 9 presents a monthly statistical analysis of the daily PM10 concentration which exceeded the guideline values for the LV (centile 90,4) during August 2007-August 2008 at BON. Importantly, at LUT, the number of days ≥ 50 µg/m^3 was null. It may be seen that the number of days exceeding the limit value for the protection of the human health (centile 90,4 : 50 µg/m^3 not to be exceeded more than 35 days/year) is of 80 days at BON (and 0 days at LUT) during the study period. Also, the average PM10 concentration during August 2007-August 2008 is 24 µg/m^3 at LUT while it is 42 µg/m^3 at BON. The results suggest that the quality objective (30 µg/m^3, on annual average) and the limit value (40 µg/m^3, on annual average) are exceeded at BON, which is not the case at LUT.

4.4. Determination of the Marine Aerosols Contribution on the PM10 Concentration

It was important to assess the chemical composition of the sampled PM10 at both LUT and BON in order to confront/comfort our assessments, concerning the origin of the sampled air at both sites. Thus, the sampled filters from the FDMS/TEOM was analysed in laboratory in order to determine the concentration of major ions present on these filters. The objective of this analysis was to determine the presence of the sea-salt (atmospheric tracers of marine origin), mainly the Chloride ions, in order to look for the marine influence on the PM10 measured at BON. The filters which were analysed in laboratory were sampled from February 2008 to April 2008 at LUT and BON.

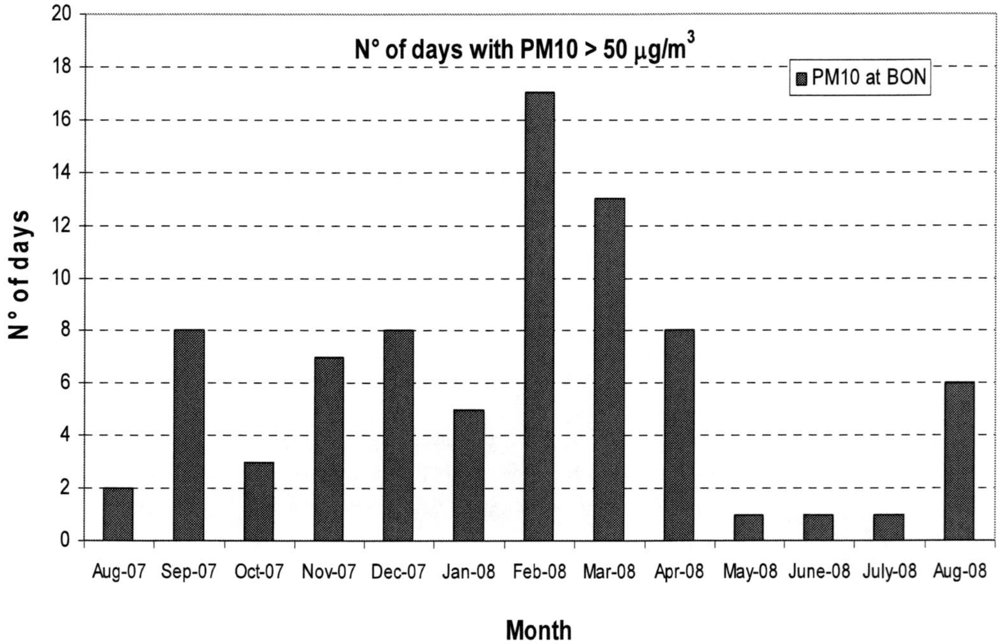

Figure 9. Monthly analysis of the daily PM10 concentration which exceeded the guideline values for the LV (centile 90,4) during August 2007-August 2008 at BON.

Figure 10 presents the map indicating the distribution of the concentration of Chloride ions (in μg/filter) measured at LUT and BON from February 2008 to April 2008. The analysis results of the sampled filters are resumed in the Table 3. Although this period encompasses events corresponding to both 'off' and 'during' swell activity, it may be seen that the Chloride concentrations are notably higher at BON than at LUT. The results confirm our previous assessments concerning the significant contribution of marine aerosols (in particular of Chlorides) on the PM10 concentration measured at BON during intense swell episodes.

Table 3. Resume of the Chloride ions concentration obtained on the sampled filters at LUT and BON from February 2008 to April 2008

Sample N°	Date Exposed	Date Finished	Cl^- (μg/m^3)	Observations
A_PM10_LUT	06/02/2008	05/03/2008	4	High swell
A_PM10_BON	26/02/2008	16/04/2008	6	moderate swell

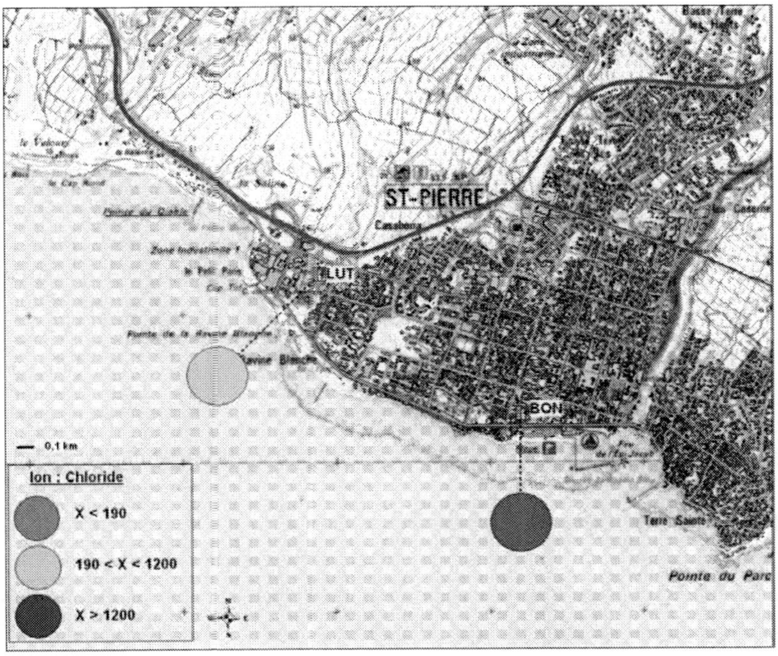

Figure 10. Map presenting the Chloride ions (in μg/filter) concentration measured at LUT and BON collected on the TEOM/FDMS filter from February 2008 to April 2008.

CONCLUSION

In this preliminary study we have characterized for the first time the PM10 concentration variability at two close urban coastal monitoring stations (distant by 1,5 km from each other): LUT and BON, situated to the south of Réunion Island, a tiny island with a population of 730 000 inhabitants, located over the south-western Indian Ocean. The aim of this work was to establish a possible link between a regional scale process such as the swell on the PM10 concentration variability at LUT and BON. As the two locations are representative of the urban pollution over the Saint Pierre agglomeration, we would expect to record similar trend and levels on the PM10 concentration profiles at both locations, which is not the case.

In this sense, we analysed the diurnal variation of the atmospheric and meteorological data in order to determine the climatology of these parameters. Then, the monthly and daily PM10 concentration variability was successively analysed at both LUT and BON, in conjunction with swell and meteorological data. The objective of these analyses was to study alternately their seasonal trends and their daily levels in comparison with guideline values and also to assess the air quality at Saint Pierre.

In order to look for the origin and main causes of the PM10 concentration variability, especially the anthropogenic contribution, nitrogen dioxide concentration monitored in parallel with PM10 at LUT and BON was also analysed. The mean diurnal PM10 concentration variation showed that the PM10 level at BON is systematically higher than at LUT by ~25 $\mu g/m^3$ during day-time and by ~6 $\mu g/m^3$ during night-time. The mean diurnal NO_2 concentration variation recorded at both LUT and BON is attributable to the anthropogenic activities (mainly traffic circulation) over Saint Pierre. The difference in the

NO_2 and the PM10 variability observed at LUT and BON point out that distinct particle sources might influence the measurement locations. The rainfall data analysis indicates that this parameter has minor influence on the PM10 concentration variability at LUT and BON, as they are influenced similarly by this parameter. The mean diurnal winds data showed that at day-time the air samples with high wind speeds were mainly of south-eastern to south-western (marine at BON and urban at LUT) origin. At night-time, the air samples with moderate wind speeds were mainly of northern to north-eastern (urban/inland) origin. The mean diurnal swell recorded at PO and PO showed a flat diurnal profile suggesting a quite continuous and constant activity throughout the day. The monthly swell level calculated in parallel with the PM10 at LUT and BON showed a good correlation mainly at BON.

The results show that daily PM10 trend and pattern is different at both sites. There is a good correlation between the daily PM10 measured at BON and the daily swell measured at PO, pointing out the potential influence of this parameter on the PM10 concentration variability at BON.

Due to the conjugate effect of the easterlies and the important relief of the island, the marine aerosols generated by the waves might also be transported towards the atmospheric monitoring stations, which explain the notable and systematic PM10 enhancement observed at BON which is situated downwind and closer to the sea than LUT. The results show that when the swell activity is important regionally, it might significantly influence the PM10 concentration measured at BON.

The LUT and BON PM10 concentrations were analysed in conjunction with winds data measured at Saint Pierre, in order to determine the origin of air masses attaining the monitoring sites.

The meteorological data analysed during the study period point out that the PM10 concentrations at BON are mainly influenced by marine air masses and confirm the potential influence of the swell activity on the PM10 variability mainly at BON. The PM10 winds rose showed that the swell regularly influence the PM10 concentration via a contribution of marine aerosols/sea spray, under certain particular conditions : throughout strong swell episodes, during sea breezes coupled with easterlies (mix of crustal/urban and marine particles) and a marine origin of the air samples with high wind speeds. The latter phenomenon contributes to systematically enhance the background PM10 levels at BON as compared to those measured at LUT.

The daily PM10 concentration recorded at LUT and BON was compared with the guideline values (centile 90,4 and annual average). Importantly, the number of days exceeding the limit value for the protection of the human health is of 80 days at BON and 0 days at LUT during August 2007-August 2008 period. The average PM10 concentration during the study period is 26 $\mu g/m^3$ at LUT while it is 42 $\mu g/m^3$ at BON. Importantly, the quality objective, the limit value and also the centile 90,4 are exceeded at BON, during the study period, while at LUT no guideline values are exceeded.

We further tried to determine the origin of the particles, via the analysis of the composition of the sampled air. In this sense, the sampled filters from the TEOM/FDMS were analysed in laboratory in order to determine the presence of Chloride ions (atmospheric tracer of marine origin). We found that the Chloride concentrations are notably higher at BON than at LUT. The results confirm our previous assessments about the significant contribution of marine aerosols on the PM10 concentration recorded at BON during intense swell and waves.

Finally, this preliminary study points out the potential of in-situ measurements to assess the origin and causes of PM10 concentration variability, in response with local-to-regional scale processes (eg. sea spray and swell activity) under 'particular' geographical (uneven relief) and meteorological conditions (land/sea breezes, winds, etc.). It constitutes an adequate tool for decision makers to take necessary steps, especially to inform the surrounding population, in order to avoid sanitary impacts, when QO and LV for human health are exceeded. However, it requires further continuous monitoring of other atmospheric pollutants (ex. marine aerosols and SO_2) 'during' and 'off' other significant swell episodes at other inhabited regions located close to the sea in order to confirm these results.

ACKNOWLEDGMENTS

We acknowledge the French 'Ministry of Environment' (MEDDTL), ADEME, CINOR, TCO and CIVIS, for their financial support of the ORA air quality network at Réunion Island.

We gratefully acknowledge the ORA staff for technical support and for the atmospheric data sampling. We thank the DDE of Réunion Island for supplying the swell data. Finally, Météo-France is gratefully acknowledged for providing the meteorological data.

REFERENCES

ADMINET, Decree n° 2002-213 of 15th February 2002 (In French), http://www.droit.org/jo/20020219/ATEP0190081D.html (2002). 25th August 2008, 2002.

Andreae, M.O., Merlet, P. Emission of trace gases and aerosols from biomass burning, *Global Biogeochemical Cycles*, Vol. 15, N° 4.. 955-966, DOI: 10.1029/2000GB001382, 2001.

ATSDR. Agency for Toxic Substances and Disease Registry, Department of Health and Human Services, *Air*, 2008.

Bhugwant C, Cachier, H., Bessafi, M., Leveau, J. Impact of traffic on black carbon aerosol concentration at La Reunion Island (southern Indian Ocean). *Atmos. Environ.*, 34, 3464-73, 2000.

Bhugwant, C., Brémaud, P. Simultaneous Measurements of Black Carbon, PM10, Ozone and NOx Variability at a Locally Polluted Island in the Southern Tropics. *Journal of Atmos. Chem.*, Vol. 39, N° 3, Pages 261-280, 2001.

Bhugwant, C., Siéja, B., Bessafi, M., Staudacher, T., Ecormier, J. Atmospheric sulfur dioxide measurements during the 2005 and 2007 eruptions of the Piton de La Fournaise volcano: Implications for human health and Environment changes, *J. Volcanol. Geotherm. Res.*, 184, 208-224, 2009.

Brunekreef B, Beelen R, Hoek G, Schouten L, Bausch-Goldbohm S, Fischer P, Armstrong B, Hughes E, Jerrett M, van den Brandt P. Effects of long-term exposure to traffic-related air pollution on respiratory and cardiovascular mortality in the Netherlands: the NLCS-AIR study, *Res Rep Health Eff Inst.*, 139, 5-71, 73-89, 2009.

Byrne, M.A., Jennings, S.G. Scavenging of sub-micrometer aerosol particles by water drops, *Atmos. Environ.*, 27A, 14, 2099-2105, 1993.

Cheng, W.L.: Ozone distribution in coastal central Taiwan under sea-breeze conditions. *Atmos. Environ.*, vol. 36, 3445-3459, 2002.

Council Directive, 1999/30/EC of 22 April 1999 relating to limit values for sulphur dioxide, nitrogen dioxide and oxides of nitrogen, particulate matter and lead in ambient air, Official Journal, L 163, 29/06/1999, 0041-0060, 1999.

Crutzen P, Andreae M.O., Biomass Burning in the Tropics: Impact on Atmospheric Chemistry and Biogeochemical Cycles, Science, 250, 4988, 1669-1678, 1990. DOI: 10.1126/science.250.4988.1669

De Leeuw, G., Vertical profiles of giant particles close above the sea surface, *Tellus, 38B,* 51-6I, 1986.

De Leeuw, G., Andreas E.L., Anguelova, M.D., Fairall, C.W., Lewis, E.R., O'Dowd, C. , Schulz, M. and Schwartz, S.E. Production flux of sea spray aerosol, *Rev. Geophys*., 49, RG2001, 2011. DOI:10.1029/2010RG000349

Dockery D.W., Pope, C.A., Acute respiratory effects of particulate air pollution, *Annual Reviews of Public Health,* 15, 107-132, 1994.

Don Porto Carero, A., P. Hoet, H.M., Verschaeve, L., Schoeters, G., Nemery, B. Genotoxic effects of carbon black particles, diesel exhaust particles, and urban air particulates and their extracts on a human alveolar epithelial cell line (A549) and a human monocytic cell line (THP-1), *Environmental and Molecular Mutagenesis*, Vol. 37, Issue 2, 155-163, 2001.

EPA, Fourth External Review Draft of Air Quality Criteria for Particulate Matter, June 2003.

Environnement Magazine, *Chantiers à risques calculés* (in French), n°1665, March 2008.

EC-COM, Communication from the Commission to the European Parliament, Thematic Strategy on air pollution, COM(2005), 446, 2005.

Finlayson-Pitts B., Pitts, J.N., Jr. Chemistry of the Upper and Lower Atmosphere, Theory, Experiments and Applications, Ed. B. Finlayson-Pitts & J. N. Pitts Jr., Academic Press, 1999.

Franklin M, Zeka A, Schwartz J. Association between PM 2.5 and all-cause and specific-cause mortality in 27 US communities. *J Expo Sci Envin Epidemiol.*, 17:279-287, 2007.

Friedlander S K., Smoke, dust and haze : fundamentals of aerosol dynamics, Oxford Univ Pr, 2000.

Goldammer, J. G. The wildland fire season 2002 in the Russian Federation: An assessment by the Global Fire Monitoring Center (GFMC), International Forest Fire News (IFFN), 28, 2-14, 2003.

Gehring U, Heinrich J, Krämer U, Grote V, Hochadel M, Sugiri D, Kraft M, Rauchfuss K, Eberwein HG, Wichmann H-E. Long-term exposure to ambient air pollution and cardiopulmonary mortality in women. *Epidemiology*, 17, 545-551, 2006.

INSEE. TER (Tableau Economique de La Réunion), (In French), 3014, 2010.

Johansson, C., Norman, M., Gidhagen, L. Spatial & temporal variations of PM10 and particle number concentrations in urban air, *Environmental Monitoring and Assessment*, Vol. 127, N° 1-3, 477-487(11), 2007.

Katsouyanni, K. Ambient air pollution and health, British Medical Bulletin, 68, 143-156, 2003.

Laden, F., Neas, L. M., Dockery, D. W., and Schwartz, J. Association of fine particulate matter from different sources with daily mortality in six US cities, *Environ. Health Perspect.*, 108, 941-947, 2000.

Liousse, C., Devaux C., Dulac, F., Cachier, H., Ageing of savanna biomass burning aerosols: Consequences on their optical properties, *Journal of Atmospheric Chemistry*, 22, 1-2, 1-17, 1995. DOI: 10.1007/BF00708178

Météo-France. Atlas climatique de La Réunion (In French). Bureau d'étude climatologique, Direction Interrégionale de La Réunion. *Annual Report*, N° 1657, 2000.

Niemi, J.V., H. Tervahattu, H. Vehkam¨aki, J. Martikainen, L. Laakso, M. Kulmala, P. Aarnio, T. Koskentalo, M. Sillanp¨a¨a, U. Makkonen. Characterization of aerosol particle episodes in Finland caused by wildfires in Eastern Europe, *Atmos. Chem. Phys.*, 5, 2299-2310, 2005.

Okada, K. Nature of individual hygroscopic particles in the urban atmosphere, *Journal of the Meteorological Society of Japan*, vol. 61, n°5, 727-736, 1983.

O'Neill, M.S., Loomis, D., Meza, V.T., Retama, A., Gold, D. Estimating particle exposure in the Mexico City metropolitan area, *Journal of Exposure Analysis and Environmental Epidemiology*, 12, 145–156 10.1038/sj/jea/7500212, 2002.

Osthoff, H.D., J.M. Roberts, A.R. Ravishankara, E.J. Williams, B.M. Lerner, R. Sommariva, T.S. Bates, D. Coffman, P.K. Quinn, J.E. Dibb, H. Stark, J.B. Burkholder, R.K. Talukdar, J. Meagher, F.C. Fehsenfeld & S.S. Brown, High levels of nitryl chloride in the polluted subtropical marine boundary layer, Nature Geoscience, 1, 324-328, 2008. doi:10.1038 /ngeo177.

Penttinen, P., Timonen, K.L., Tiittanen, P., Mirme, A., Ruuskanen J., Pekkanen, J. Ultrafine particles in urban air and respiratory health among adult asthmatics, *Eur Respir J.*; 17: 428-435, 2001.

Pope, C.A., Burnett, R.T., Thun, M.J., Calle, E.E., Krewski, D., Ito, K., Thurston G.D. Lung cancer, cardiopulmonary mortality, and long-term exposure to fine particulate air pollution, *JAMA-Journal of the American Medical Association*, 288(7):830, 2002.

Pope CA, Ezzoti M, Dockery DW. Fine Particulate Air Pollution and Life Expectancy in the United States. *N. England J. Med.*, 360, 376-86, 2009.

Querol, X., A. Alastuey, S. Rodriguez, F. Plana, C. R. Ruiz, N. Cots, G. Massagué, O. Puig. PM10 and PM2.5 source apportionment in the Barcelona Metropolitan area, Catalonia, Spain, *Atmospheric Environment*, 35, 36, 6407-6419, 2001.

Querol X., Pey, J., Pandolfi, M., Alastuey, A., Cusack, M., Pe´ rez, N., Moreno, T., Viana, M., Mihalopoulos, N., Kallos, G., Kleanthous, S. African dust contributions to mean ambient PM10 mass-levels across the Mediterranean Basin, *Atmospheric Environment*, 43, 4266-4277, 2009.

Rabl, A., Teulère, L. Estimation des coûts attribuables à la pollution de l'air dans le secteur du bâtiment, *Pollution Atmosphérique* (in French), N° 164, 81-91, 1999.

Ramanathan, V., P. J. Crutzen, A. P. Mitra, and D. Sikka, 2002. The Indian Ocean experiment and the Asian brown cloud, *Current Science*, 83(8), 947-955.

Ruidavets, J.B., Cournot, M., Cassadou, S., Giroux, M., Meybeck, M., Ferrières, J. Ozone air pollution is associated with acute myocardial infarction, Circulation. American Heart Association, Inc., 111: 563-569, 2005.

Samet, J. M., F. Dominici, F. C. Curriero, I. Coursac, and S. L. Zeger, Fine Particulate Air Pollution and Mortality in 20 U.S. Cities, 1987-1994, *New England J. Med.*, 343:1742-1749, 2000.

Seinfeld J.H., Bassett M. Effect of the gas-to-particle conversion on the evolution of aerosol size distributions, in Heterogeneous atmospheric chemistry, D. R. Schryer Ed., *Geophysical monograph*, 26, 6-12, 1982.

Stull, R.B. An Introduction to Boundary Layer Meteorology, Springer Publishing, New York, 150-152, 1988.

Uematsu, M., K. Kinoshita K., Nojiri, Y. Scavenging of Insoluble Particles from the Marine Atmosphere over the Sub-Arctic North Pacific, *Journal of Atmospheric Chemistry*, 35, 2, 151-163, 2000.

Viana, M., Pérez, C., Querol, X., Alastuey, A., Nickovic, S., Baldasano, J.M. Spatial and temporal variability ofPM levels and composition in a complex summer atmospheric scenario in Barcelona (NE Spain), *Atmospheric Environment*, 39, 5343-5361, 2005. doi:10.1016/j.atmosenv.2005.05.039

Wanjura, J.D., B. Shaw, Jr. C. Parnell, R. Laey, R., S. Capareda. Comparison of continuous monitor (TEOM) and gravimetric sampler particulate matter concentrations, *Transactions of the ASABE*, 51(1):251-257, 2008.

Weijers, E.P., Schaap, M., Nguyen, L., Matthijsen, J., Denier van der Gon, H.A.C., Ten Brink, H.M., Hoogerbrugge, R. Anthropogenic and natural constituents in particulate matter in the Netherlands, *Atmos. Chem. Phys.*, 11, 2281-2294, 2011. doi:10.5194/acp-11-2281-2011

WHO. Health Aspects of Air Pollution with Particulate Matter, Ozone and Nitrogen Dioxide", *Report on a WHO Working Group*, 2003.

WICKS, P. Clean air for Europe, *Pollution atmosphérique* (in French), n° Mai 2002, 7-14, 2002.

In: Air Pollution Emissions
Editors: Daniel G. Vasiliev and Robert A. Kask
ISBN: 978-1-62100-453-0

Chapter 4

INVENTORY AND BEHAVIOUR OF AIR POLLUTANTS IN AN URBAN AREA OF THE NW OF SPAIN

Jorge Sanjurjo-Sánchez*
University Institute of Geology 'Isidro Parga Pondal',
University of A Coruña, Campus de Elviña, 15071 A Coruña, Spain

ABSTRACT

Directives for evaluating the ambient air quality in the EU have been adopted in the last years as a part of the new strategies for pollution prevention and control. Such strategies are based in the control of emissions and measurement of gaseous and particulate pollutants in the air, in industrial, urban and rural areas. Inventories of industrial emissions and concentrations measured in control stations of NO_2, NO, O_3, SO_2, CO, CO_2 and particulate matter have been monitored in urban areas of the EU. As a result, the knowledge of dispersion, interaction and general behaviour of such pollutants have increased. Both the dispersion of pollutants, and daily, weekly, sesonal and annual cycles caused by the emissicns have been studied in different areas or the World by using such control stations. In the NW of Spain, available data from emissions and measurements of pollutants in the stations have provided valuable information about the behaviour of the pollutants in the urban area of A Coruña city. In this area the behaviour of pollutants is marked by the seaside climate. The data compiled in the year 2007 are presented here. They have allowed understand the main sources causing dangerous concentrations of pollutants, to use such information for future urban planning. Results of such observations indicate that near sources of air pollutants can be responsible of mean and peak concentrations. Also, the trend of some pollutants can be related to climatic factors.

Keywords: Air pollution, urban settings, decay, stone materials

* Email: jsanjurjo@udc.es.

1. Introduction

Both gaseous compounds and particulate matter are emitted to the air by anthropogenic activities and contribute to the deterioration of air quality. This has encouraged diverse studies on the chemistry, distribution and effects of gaseous and particulate pollutants in urban areas. Inventories of industrial emissions and concentrations measured in control stations of gaseous NO_2, NO, O_3, SO_2, CO, CO_2 and particulate matter have been monitored in such areas in the last decade. Current EU legislation controls the concentration of such pollutants by imposing limit values [1, 2, 3]. The control is based in the real-time measurement of concentration of pollutants near industrial and urban areas. Also, annual industrial emissions must be declared [4].

Atmospheric chemical constituents can precipitate by wet occult and dry deposition. In humid climates, where important rainfalls are recorded, wet deposition must be considered as an important source of ions and particles, and no doubt exists about acid rain damage on soils, plants, animals, humans and even building materials. Occult deposition or precipitation refers to precipitation of cloud/fog water. It can be more damaging when compared to dry and wet deposition, providing moisture to activate the chemical attack of dry deposited gases and particles. Fog water has been shown to be an important collector and carrier of pollutants in urban areas as it can concentrate chlorides and sulphates originated from local emission sources [5]. At seaside areas, seawater sulphate and salts due to sea spray may contribute to the wet and occult deposition [6]. Dry deposition removes gaseous and particulate pollutants from the atmosphere in the absence of precipitation [7]. It is determined by the concentration of pollutant gases and particles in the atmosphere, and the interaction between the atmosphere and surfaces under dry conditions. The dilution and transport of pollutants, including PM, are controlled by atmospheric transport, dispersion and removal mechanisms, depending on topographic features and the air and climatic conditions of an area.

Adverse health effects of air pollutants have been widely studied and described. Among them particles have received particular attention. Toxicity of PM depends on their composition and also their size: particulate matter of grain size less than 10 μm and 2.5 μm are known as PM_{10} and $PM_{2.5}$, respectively. The studied health endpoints in relationship with PM are mortality, hospital admissions due to cardiovascular and respiratory diseases and lung function among others [8]. Gaseous compounds have also been linked to some health problems, namely SO_2-CO and O_3 have been related with asthma hospitalizations and asthma-wheezing emergency room visits, respectively. Nitrogen oxides have been related with diverse respiratory hospital admissions, asthma and following respiratory infections [9].

The main gaseous pollutants resulting from anthropogenic sources are presented in table 1. Among gaseous pollutants, nitrogen and sulphur oxides, carbon oxides and ozone are those of main concern. They are mainly produced and emitted due to industrial activities and traffic exhaust, although important natural sources also exist. Thus, natural sources overpass anthropogenic ones, but human activities are concentrated in urban areas, where their effect can be important.

Table 1. Yearly emissions of gaseous compounds

Emission	Natural (million T)	Anthropogenic (million T)
CO_2	600,000	22,000
CO	3,800	550
Hydrocarbons	2,600	90
CH_4	1,600	110
NH_3	1,200	7
NO_2	770	53
SO_2	20	150

From Fellenberg 2000 [10].

The evaluation of historical trends of pollution in industrialized nations is useful in showing recent air quality improvements and also to understand what can be the result in air pollutant controls currently experiencing high levels of pollution. Current Spanish and Galician legislation is based on EU legislation, which controls the mass concentration of some typical gaseous compounds and PM_{10} (particulate matter of grain size less than 10 μm). Two limit values have been imposed for PM_{10}: a 2 h mean concentration of 50 μg/m^3 not to be exceeded more than 35 times during a calendar year, and an annual mean concentration of 20 μg/m^3 [1,2,3]. Moreover, PM_{10} objectives by 2010 limit the 2 h mean concentration of 50 μg/m^3, not to be exceeded more than 7 times during a calendar year. The SO_2 concentration is limited to a 1 h mean concentration of 350 μg/m^3 (it cannot be exceeded more than 24 times), 24 h mean concentration of 125 μg/m^3 (3 times) and annual mean concentration of 20 μg/m^3. The NO_2 concentration is also limited to a 1 h mean concentration of 200 μg/m^3 (exceeded 18 times) and an annual mean concentration of 40 μg/m^3, while NO_x cannot overpass an annual mean concentration of 30 μg/m^3. The mean concentration of CO cannot overpass 40 μg/m^3 for 8 h in a day.

1.1. Gaseous Compounds

Gaseous sulphur is naturally produced by volcanic activity, sea spray and different biogenic processes. Anthropogenic come from combustion processes: coal combustion in thermal power plants and refining processes of crude oil [11]. Sulphur gaseous compounds are mainly present in the lower troposphere as sulphur dioxide (SO_2), sulphur trioxide (SO_3) and hydrogen sulphide (H_2S).

Hydrogen sulphide is an occasional atmospheric pollutant as it is quickly oxidized [12]. Sulphur oxides and particularly SO_2 are considered as the most important gaseous compound. An unsolved question is whether sulphites are formed and subsequently oxidized or whether sulphates are the only reaction products. The oxidation of SO_2 can occur in the gas phase or in the gas-liquid or gas-solid phase [12,13,14]. In the gaseous phase, the reaction is a homogeneous process of oxidation of SO_2 in air that involves photo-excitation and absorption of solar radiation. This process is strongly influenced by low concentrations of nitrogen oxides and some hydrocarbons in polluted air [12,15].

Natural sources of nitrogen oxides (NO and NO_2, usually referred as NO_x) include volcanoes, oceans, biological decay and lightning strikes. Vehicle engines are the main anthropogenic source of nitrogen oxides in the urban environment. Emissions from industrial combustion of coal in coal-fired power plants, boilers, warm air furnaces, gas turbines, open burning, structural fires, explosives and fertilizers must be considered [16,17]. Also, nitrous oxide (N_2O) is emitted by natural sources (tropical soils, near-surface oceanic emissions) and anthropogenic activities.

In urban settings, gasoline and diesel engines are major sources of NO and smaller amounts of NO_2. Diesel engines produce five times more NO_x. At high relative humidity, NO is very rapidly oxidised to NO_2 [11]. Directly emitted NO_2, also called primary NO_2, are about 5% of NO_x emissions by volume [18]. Studies near major highways show NO_2 concentrations positively correlated with traffic density [19,20]. Although the total emissions of nitrogen oxides in Europe have declined in the last years [21], other factors may counteract the effect of these reductions. There is strong evidence of recent increases in the fractional contribution of NO_2 to NO_x emissions from vehicle emissions in Europe, to levels around 10% around 2000 and up to 20% in 2005 in the UK [22].

NO_x strongly react in the presence of sunlight by way of a series of photochemical reactions [11] with ozone (O_3) to form secondary NO_2. Although a large number of chemical processes potentially oxidize NO to NO_2 in the atmosphere [23], the most important source of secondary NO_2 is the slow reaction of NO with O_3 to give NO_3 during the diurnal cycle [23].

Background O_3 has been rising in the last decades [22], although not at a constant rate, lead to a faster oxidation of NO to NO_2, enhancing the NO_2 concentration [24]. Nitrogen dioxide (NO_2) and nitric oxide (NO) play a key role in determining concentrations of O_3, nitric acid (HNO_3), nitrous acid (HNO_2), organic nitrates, nitrate aerosols and other species in the troposphere [25]. The photolysis of NO_2 in the presence of volatile organic compounds (VOCs) is considered as the only key initiator of the photochemical formation of O_3 and photochemical smog [26]. VOCs are a major group of pollutants [27] that undergo complex photochemical reactions in the lower troposphere. On a regional scale a large number of reactions, including VOCs and CO, are important to determine the O_3 concentration, while on a local scale the dynamics of NO-NO_2-O_3 is dominated by a limited number of fast reactions [24].

Although O_3 can be produced by natural sources (e.g. trees, thunderstorms), episodes of high O_3 concentration are frequent in densely populated areas, related to forest decline and industrial emissions [28]. Atmospheric O_3 in the lower troposphere is produced by photolysis of nitrogen dioxide (NO_2) and subsequent recombination between atomic and molecular oxygen in the presence of a third-body molecule. VOC in the atmosphere interrupt this process by being more attractive to the NO that the O_3 molecule, contributing to O_3 accumulation [29,30,31]. The photochemical O_3 production in urban areas is less sensitive to the VOCs emissions [32]. An enormous variety of VOC classes may be emitted from numerous anthropogenic and biogenic sources [33,34] and, depending on location, either or both categories can make a major contribution to photochemical O_3 formation [23,35]. In areas moderately contaminated, O_3 sensitivity to emission of nitrogen oxides depends on the season and on the emission rates [36]. The non-linearity in the NO_2 vs. NO_x concentration relationship results mainly from the chemical coupling of O_3 and NO_x and any reduction in the NO_2 concentration is accompanied by an increase in the O_3 concentration [18,37,38].

Carbon oxides (monoxide and dioxide) are normal constituents of the atmosphere, as a part of the exogenous carbon cycle. CO_2 concentrations are highly variably depending on the area [12]. Natural sources of carbon dioxide in the atmosphere include combustion of organic matter, respiration processes in organisms and volcanic outgassing. Anthropogenic sources are related to the burning of fossil fuels for power generation, transport and heating and other industrial processes. Thus, their concentration can be higher in urban areas. The concentration of carbon dioxide in the Earth's atmosphere has been rising 1.9 ppm/yr in average during the period 2000-2008 [39]. Estimated emissions of traffic exhaust are responsible by 10% of the CO_2 global emissions [11].

Carbon monoxide (CO) exists in lower concentration than CO_2 in air. Natural sources of CO are natural gases, volcanic outgassing, forest fires, bacterial activity and methane oxidation in the atmosphere. Anthropogenic CO is a final combustion product if the amount of O_2 is not sufficient for complete oxidation, sometimes due to a bad mixing of the air and fuel. In urban areas CO concentrations can reach 100 ppm and higher [10]. It is considered that about 25% of CO in the atmosphere was produced by direct anthropogenic emissions [12].

1.2. Particulate Matter

The term atmospheric particulate matter is used to refer fine solid or liquid particles suspended in air. Particulate matter (PM) is important for human health and other animals, ecosystems and processes related with environmental pollution and atmospheric chemistry. They also have environmental effects including changes in visibility, solar radiation transfer (related with global warming), cloud formation, and play a major role in the acidification of clouds, rain and fog [40].

Atmospheric particles can be classified considering their size. The largest particles are short-lived and remain airborne near to their source due to their high rate of sedimentation, while the smallest ones have very short lifetimes in air because their attachment to larger particles. They tend to be modified with time due to coagulation, adsorption of gases, dry and wet removal and chemical reaction within particles in the presence of liquid water [12].

Atmospheric particles may be originated by natural and anthropogenic sources. Natural sources o are volcanic outgassing, forest fires, sea salt and gas-phase conversion of other atmospheric compounds. The main anthropogenic sources are burning of fossil fuels (industrial, transport and domestic burning), diverse industrial processes, mining and agriculture. Due to climatic factors, the cyclic human activity and the diurnal cycles, the concentration of suspended particulate matter in the air oscillates diary, weakly and yearly [10,12]. In urban settings fossil fuel combustion by industrial facilities and vehicular engines can be responsible by high concentration of particles in the air. However, it is very difficult to assess the sources (natural or anthropogenic) of atmospheric particles from their composition.

The distribution of atmospheric particles in urban settings will depend on the characteristics of the urban planning. In canyon streets higher PM emitted by car engines have been found closer to the ground [41]. Four main causes of particulate pollution episodes have been observed in European cities [42]: strong traffic-related emission sources, atmospheric local dispersion conditions, synoptic weather conditions that favour long-range transport of particles, and natural sources of PM not easily controllable (e.g. windblown dust).

PM is important for human health. Thus, PM has been classified as PM_{10} (<10 μm in diameter) and $PM_{2.5}$ (diameter <2.5 μm) particles, considered as the most active component in human health impacts [43]. PM <1 μm in diameter (also called PM_1) are considered at present in some air pollution measurements. The number of data sets containing comprehensive chemical characterization of PM_{10} is small compared to the volume of mass concentration data. Studies show that it is very difficult to assess the origin sources (natural of anthropogenic) of PM [44,45,46,47,48]. They show that the major components that account for nearly all of the PM_{10} mass in most urban areas includes [11,49,50,51,52,53,54,55] geological materials (oxides of Al, Si, Ca, Si, Fe, K and Ti), diverse organic compounds, elemental carbon, inorganic particulate matter, sulphates and nitrogen compounds. Wu et al. [56] concluded that the concentration of PM_{10} in air, when emitted by car engines in urban streets, decreases between 60 and 80% of the maximum occurring at 2 m above the ground. Although the highest PM_{10} concentrations generally occur at the roadside, certain urban background monitoring stations have also recorded relatively high PM_{10} values. That raises concern about the exposure of a large proportion of the European urban population to particulate pollution [42].

Regarding the chemical composition of particles can be attributed to both natural and sources as sea spray in coastal areas or wind-blow of crustal dust. This is the case of some sulphur-rich, Cl, Mg, K, Na, Ti, Al, Fe, etc. Other components have been liked to anthropogenic pollution sources such as oil combustion (Ni or V), urban solid waste (carbonaceous particles, Pb and Zn), industrial burning of coal or other industrial emissions (As, Cr, Cd, Pb, V, Zn, Sn, Cu, B and carbonaceous particles) and gasoline and diesel combustion (sulphur- and nitrogen-rich salts, Pb, Cl, Br, Cd, Zn, Al, Si, carbonaceous particles, volatile compounds and fly ash) [50,51,52,53,54,55,57,58].

The combustion of fuel in transport is a significant source of carbonaceous particles (carbon soot or fly ash by incomplete combustion) and metallic particles such as Al, Si, Zn, Cd, Pb [11,49,59]. Other particulate emissions caused by car traffic such as brake dust (SO_2, Mg, Fe, Si, carbonaceous particles and organic carbon), tire dust (SO_2, Fe, Si, carbonaceous particles and organic carbon) or paved road dust (Si, Al, K, Ca, Ti, Fe, SO_2, carbonaceous particles and organic carbon) are not usually considered in pollution studies [49].

Human activities in urban areas also contribute to particle emissions through [49] oil boilers (SO_2, NH_3 and carbonaceous particles), wood fires (Cl, carbonaceous particles and organic carbon), synthetic fires (Cl, Al, SO_2, carbonaceous particles and organic carbon), natural gas home appliances (SO_2, NO_3, Cl, Na, Ca, carbonaceous particles and organic carbon), meat cooking (Cl, Na, SO, NO_3, Cu and organic carbon) or vegetative detritus (Si, Al, Cl, K, Ca, Fe, Cu, Zn and organic carbon) [49].

2. Study Case

The aim of this work is to compile the atmospheric emissions and to observe their effect in gaseous and particulate compounds measured in air during the year 2007 in the urban area of A Coruña (NW Spain). A Coruña is a city located on the NW coast of Spain, with approximately 250,000 inhabitants, but its metropolitan area reach more than 500,000. Vehicular traffic is abundant in the city and the outskirts, and several industrial facilities cause important emissions of pollutants in the area. The climate of the city is sub-humid Mediterranean with Atlantic trend. The average annual temperature is 13.9° C and the average annual rainfall 1000 mm [60]. Rainfall occurs during all the year but they are more abundant in winter. The mean temperatures slightly oscillate seasonally: 12° C in winter and 19°C in summer.

A compilation of recent data [61] of wind measurements indicates that in both places, the prevailing wind direction is S-N/SSW-NNE although the resultant mean force of the winds is S-N direction. Such wind direction is more or less constant during all the seasons of the year. In winter, strong winds are recorded. Even winds of 14-19 km/h can be recorded an average of 3-4 days per month at this time of year, most of them NE-SW sense.

Data from air pollution originated from vehicular traffic in the area are provided by studies of mean daily traffic intensity by the council of A Coruña. The estimated average daily traffic is about 100,000 vehicles. Regarding to industrial sources of air pollution (figure 1), in the outskirts of A Coruña there are a thermal power station (Sabón), an oil refinery and an aluminium production factory (situated 10 km SW), which emit high amounts of pollutant gases and particles. Other industrial activities in this location are production of silicate compounds, graphite production, meat transformation and beer production. Another thermal power station (Meirama) is located approximately 20 km S from the city centre. Ceramic industries exist located 20 km SW, and other factories of bioethanol production (35 km SE), wood transformation and meat production (20 km SE) are dispersed in the suburban area. Emissions of atmospheric gases and particles from these industrial sources have been compiled from environmental annual reports published by the government of Spain of 2007. Data reports have been used to build emissions inventories [4].

Table 2. Air quality stations used in the study, council and urban arrangement in the sites where the stations are located

Point	Station	Council	Area type
1	A Coruña	A Coruña	Urban (city centre)
2	Arteixo	Arteixo	Suburban (industrial area)
3	A Grela	A Coruña	Urban (industrial area)
4	Lañas	Arteixo	Suburban (industrial area)
5	Pastoriza	Arteixo	Suburban (industrial area)
6	Paiosaco	Arteixo	Suburban (industrial area)
7	Sorrizo	Arteixo	Suburban (near industrial area)
8	San Vicente	Cambre	Suburban (near industrial area
9	Mesón	Carral	Rural (near industrial area)
10	Monte Xalo	Carral	Rural (clean air)
11	Xalo	Carral	Rural (clean air)
12	Cerceda	Cerceda	Rural (industrial area)
13	Rodis	Cerceda	Rural (industrial area)
14	Paraxón	Cesuras	Rural (clean air)
15	Bemantes	Miño	Suburban (far from industrial areas)
16	Cendón	Laracha	Suburban (near industrial area)

Despite data from emissions can be useful to assess air quality in an urban area, air quality information systems provide real-time data of the hourly concentration of pollutants in the air. A network of air quality monitoring stations exists in the area of A Coruña (table 1), supported by the Government of Galicia (NW Spain). The stations are located near to the industrial sources of the area, in the outskirts of the city and the city centre (industrial, suburban and traffic, respectively), and areas far from the urban centre and industrial facilities. The data provided by the network allowed us to know the hourly concentration of gaseous compounds and PM_{10} in the air, to assess seasonal patterns and annual mean concentrations. Measurements from sixteen stations were used (A Coruña-city centre, Arteixo, A Grela, Lañas, Paiosaco, Pastoriza, Sorrizo, San Vicente, Mesón, Monte Xalo, Xalo, Cerceda, Rodis, Paraxón, Benantes, Cendón) to assess the air quality. The location of these stations is shown in figure 1. The comparison of the data will allows assessing the effect of industrial and traffic pollution emissions in the air quality of the whole area.

The experimental period for this study was the natural year 2007. These data were obtained by the Public Health Department of the Government of Galicia [62] and the National Institute of Meteorology. Measurements are performed by different methods: PM_{10} levels were measured (beta attenuation and oscillating micro-balance), depending on the station, measurements of are carried out by UV fluorescence SO_2, NO_2/NOx are measured by chemiluminescence, data of O_3 is obtained by UV photometry and CO is measured by non-dispersive IR.

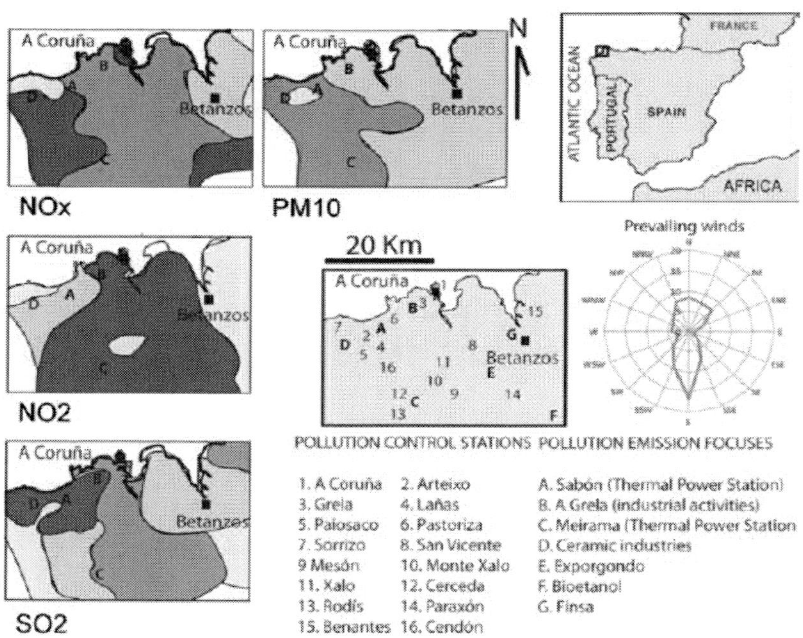

Figure 1. Map of the area of A Coruña with location of the main industrial facilities, air quality stations and mean annual concentration of some major gaseous pollutants and PM_{10}. Lighter gray indicates lower mean annual concentrations while darker gray indicates higher concentration. Data on the wind directions are provided.

3. EMISSIONS

Data from annual emissions have been compiled from reports published by the Spanish Government [4]. From this data, temporal and variability of emissions cannot be inferred. Thus, temporal oscillations or trends must be observed on data collected in air pollution stations. However, the annual bulk emissions are interesting to know the inventory of emissions and their spatial distribution to understand their effect, diffusion processes and oscillations in the urban area. Undoubtedly, the most important emission focuses are the oil refinery and the two thermal power stations located in the area (figure 2). They emit most carbon gaseous compounds (CO, CO_2 and CH_4) and sulphur oxides. They also emit nitrogen gaseous compounds, but other industrial activities emit similar amounts. Also, particulate matter is emitted due to other industrial activities in the area. The summary of the industrial emissions in the area during the year 2007 is shown in the table 3 and figure 2.

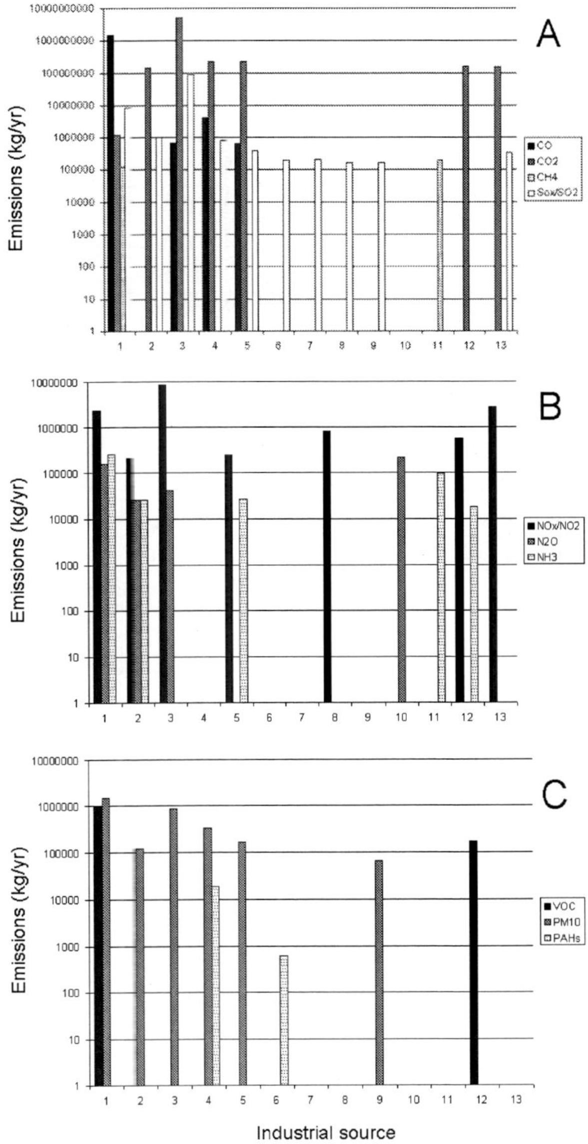

Figure 2. Histogram of industrial emissions in the area of A Coruña city in 2007. Key: industrial sources: 1, oil refinery; 2, Sabón thermal power station; 3, Meirama thermal power station; 4, aluminium production; 5, production of silicate compounds; 6, production of graphite components; 7, beer production; 8, meat transformation; 9, ceramic production; 10, ceramic production; 11, meat production; 12, bioethanol production; 13, wood transformation.

Data from air pollution originated from vehicular traffic in the area are provided by studies of mean daily traffic intensity by the council of A Coruña. The estimated average daily traffic is about 100,000 vehicles in A Coruña city and about 200,000 in the urban area. Data of gaseous and particulate emissions of vehicle engines can be inferred from published data on data compiled in the area [63,64], data on the composition of traffic emissions [11,20,49], and data of gaseous compounds and particulate matter recorded in air pollution stations.

Table 3. Name and activity of the industrial facilities in the area of A Coruña city (the number correspond with the number in figure 2

Industry	Industry name (Enterprise)	Activity
1	Repsol	Oil Refinery
2	Sabon (Unión Fenosa)	Thermal power station (gas)
3	Meirama (Unión Fenosa)	Thermal power station (coal)
4	Alcoa	Aluminium production
5	Feroatlántica Sabón	Production of silicate compounds
6	SGL Carbón	Production of graphite components
7	Hijos de Rivera SA	Beer production
8	Artabra SA	Meat transformation
9	Cerámicas El Progreso	Ceramic production
10	Campo Brick SL	Ceramic production
11	Exporgondo	Meat production
12	Bioetanol Galicia	Bioethanol production
13	Finsa	Wood transformation

4. Behaviour of Pollutants

4.1. Daily and Weekly Measurements

The temporal variability of gaseous pollutants and PM_{10} has been observed towards measurements in the city centre, as occurs in other urban areas of the world. Daily measurements of both SO_2 and CO hourly concentrations fit very well (figures 3a and 4a). Both compounds show the same pattern, indicative of similar major sources and diffusion with similar concentrations during winter and summer days. The concentration of both compounds shows a major peak from 8:00 AM to 2:00 PM and a minor peak from 19:00 to 20:00 PM. Such peaks can be observed and they are similar in winter and summer days. From these data it could be difficult to assess if the hourly pattern is due to industrial of traffic emissions, as both are significant sources of such gaseous compounds. However, such daily pattern of hourly variability matches the traffic intensity peaks. Moreover, other gaseous compounds must be considered to assess the contribution of industrial emissions to such variability.

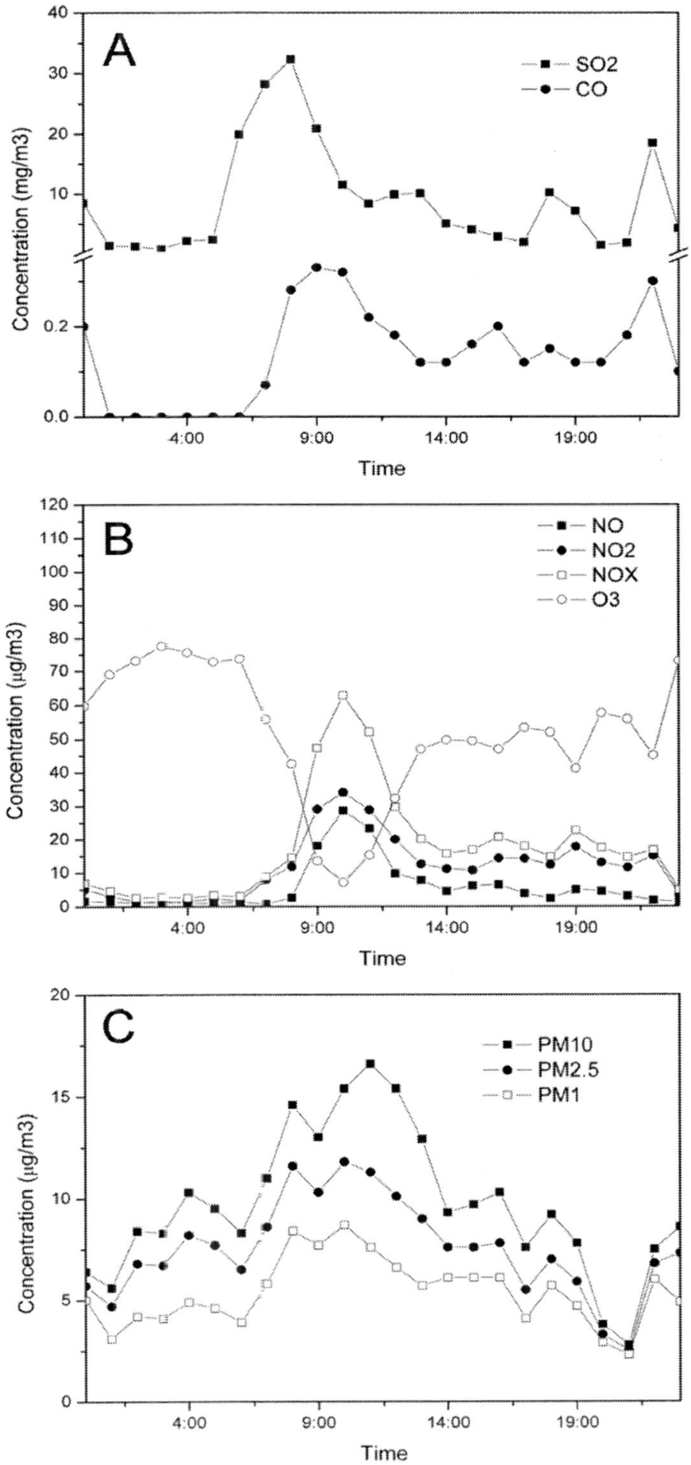

Figure 3. Daily oscillations of gaseous pollutants and particulate matter in a typical winter day: (A) SO_2 and CO, (B) NO, NO_2, NO_x and O_3 (C) PM_{10}, $PM_{2.5}$ and PM_1.

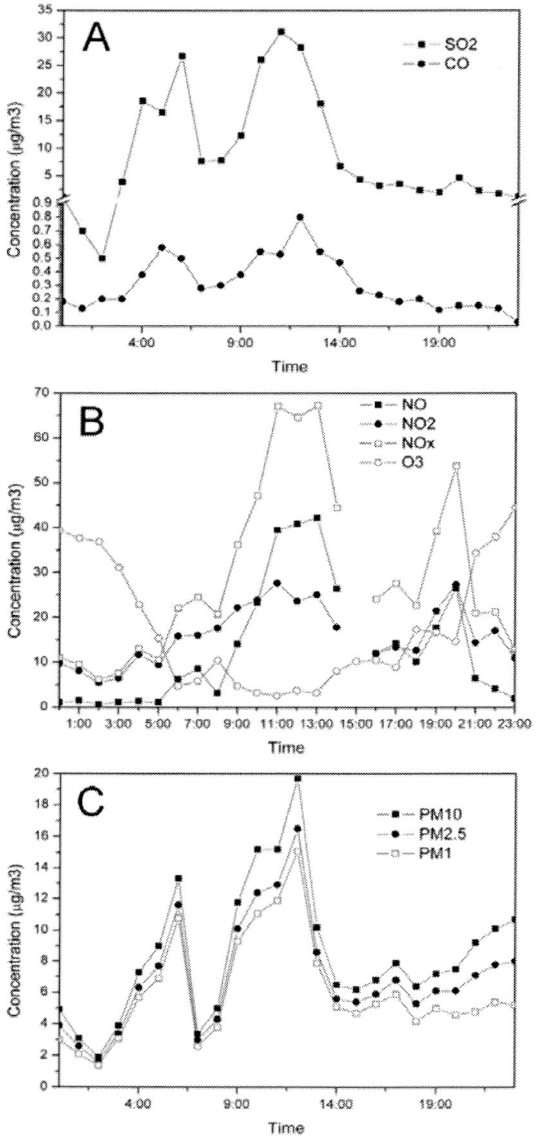

Figure 4. Daily oscillations of gaseous pollutants and particulate matter in a typical summer day: (A) SO_2 and CO, (B) NO, NO_2, NO_x and O_3 (C) PM_{10}, $PM_{2.5}$ and PM_1.

Gaseous nitrogen oxides (NO, NO_2, NO_x) show very close hourly patterns in winter and summer days and they concentration increases when O_3 decreases and vice versa. This correlation is typically observed in such gaseous compounds. The behaviour of these compounds has been studied extensively [23] and can be related with both some industrial activities and traffic emissions [11]. NO_x shows two main high peaks with similar hourly concentrations in summer and winter days: one major peak in the morning (from 8:00 AM to 2:00 PM) and another minor peak (19:00 to 20:00 PM). Such peaks fit higher traffic intensities in the urban centre (figures 3b and 4b). The combustion of diesel fuel is a major source of NO_x and thus traffic concentrated during such hours can be related with the concentration peaks [11].

Particulate matter of different size (PM_{10}, $PM_{2.5}$ and PM_1) shows very close patterns, indicative of a same origin and diffusion patterns in the urban area. The hourly concentration of particulate matter varies with a major peak in the morning, which approximately fits the gaseous peaks, although concentrations are not very different in winter and summer days. Thus, a similar pattern is observed in the city centre for hourly concentrations of gaseous compounds and particulate matter. Such variations can be related with traffic emissions if we consider three questions: traffic emissions are more important during working hours, gaseous NO_x is an important tracer of traffic emissions, and some major industrial emissions are not usually concentrated during certain hours of the day, as occurs with emissions from thermal power stations which activity strongly varies from day to day.

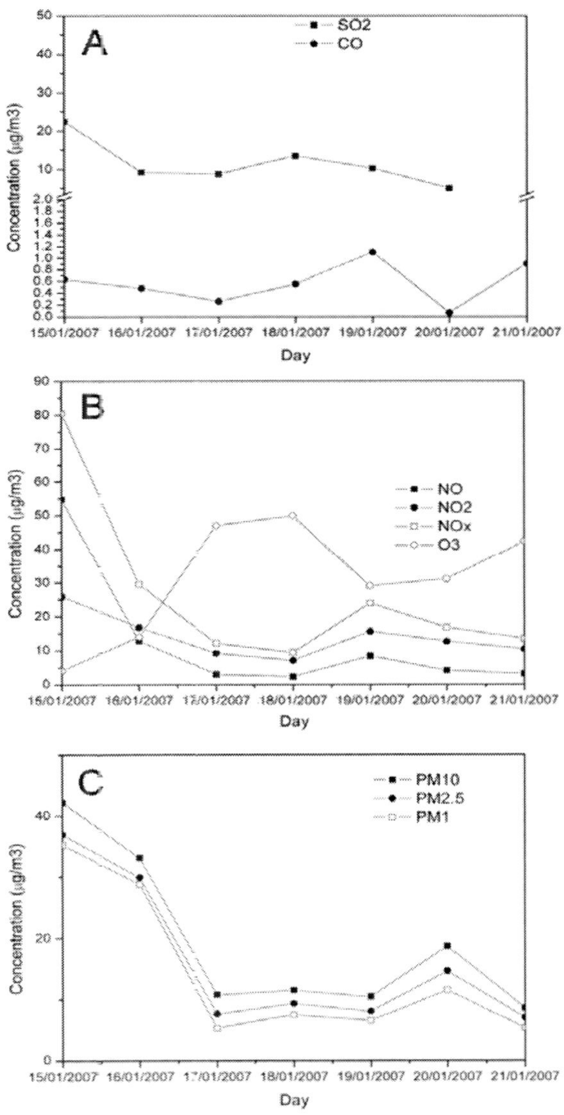

Figure 5. Weekly oscillations of gaseous pollutants and particulate matter in a typical winter week: (A) SO_2 and CO, (B) NO, NO_2, NO_x and O_3 (C) PM_{10}, $PM_{2.5}$ and PM_1.

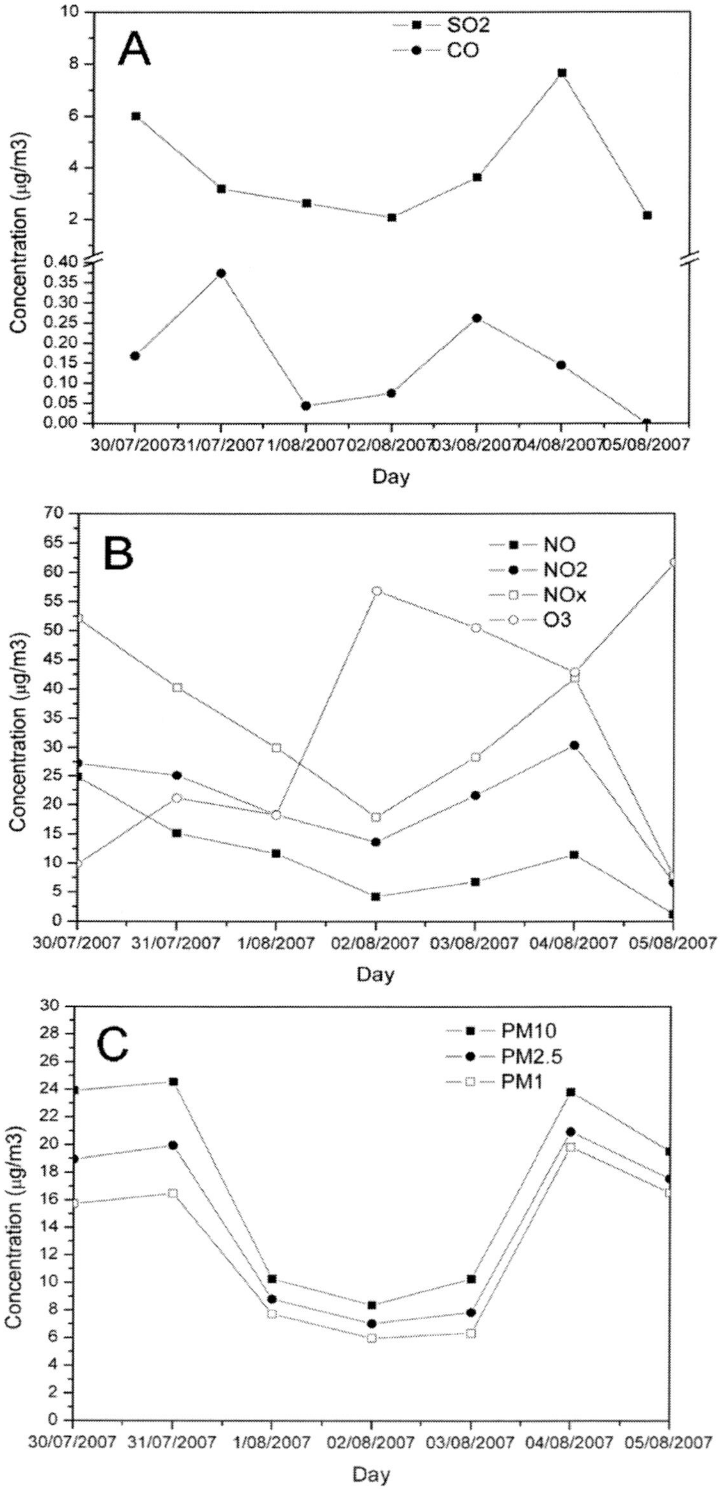

Figure 6. Weekly oscillations of gaseous pollutants and particulate matter in a typical summer week: (A) SO_2 and CO, (B) NO, NO_2, NO_x and O_3 (C) PM_{10}, $PM_{2.5}$ and PM_1.

The weekly variability in the concentration of the air pollutants is observed from their mean daily concentration. The evolution of daily averaged values of pollutants in weekly periods was observed to assess the weekly effect of industrial activities and traffic (figures 5 and 6). Oscillations on the mean SO_2 and CO daily concentration during winter and summer weeks are not important, although some small peaks can be observed on Saturdays (figures 5A and 6A). The concentration of NO_x and nitrogen gaseous compounds shows a similar pattern in winter and summer weeks, very similar to the pattern of particulate matter (figures 5B and 6B). Such patterns are very variable from day to day. Thus, not enough information can be obtained from this behaviour of the pollutants in the city centre. Particulate matter shows the same weekly behaviour during winter and summer (figures 5C and 6C). A similar weekly pattern is observed in suburban (industrial and non-industrial) and rural areas of the city outskirts for the same pollutants. Thus, the main agents responsible of oscillations of the concentration of air pollutants in the urban area are weather and traffic emissions. Industrial emissions seem play a secondary role in the area.

4.2. Seasonal Measurements

The concentration of air pollutants in the pollution stations of the urban area have been measured during 2007, although available data also exist from most stations, from 2002 to 2010. Although the levels of gaseous pollutants and particulate matter are relatively low in every time of the year, some high concentration events can be observed. For SO_2 concentration, values above 100 $\mu g/m^3$ have been observed 9 times in 2007 although the maximum legal thresholds (1 h mean concentration of 350 $\mu g/m^3$, 24 h mean concentration of 125 $\mu g/m^3$) have never been exceeded (figure 7A). Either no or small seasonal variability is observed: slightly lower concentration values are observed during summer. This trend could be related with burning of fossil fuel in thermal power stations due to domestic heating systems. However, uncertainty in the measurements of hourly and mean concentrations hinders the possibility of verifying this statement. The annual mean concentration is also below the legal threshold (20 $\mu g/m^3$). The CO concentration shows concentrations values that usually fit the SO_2 (figure 7B). Seasonal variability is not observed or is minimum, as occurs with SO_2. Both gaseous compounds are probably due to a same origin, and they show similar dilution, diffusion and behaviour depending on atmospheric conditions. The CO thresholds (40 $\mu g/m^3$) have never been exceeded in 2007 and maximum hourly values are below 4 $\mu g/m^3$ (10 times lower than legal threshold). The SO_2 and CO concentrations in other stations of the urban, suburban and closed rural areas are lower or similar.

The NO_2, NO and NO_x concentration show a different annual pattern when compared with both SO_2 and CO. Seasonal variability can be observed for nitrogen oxides. Mean concentration values during summer-autumn are slightly higher as occurs with peak concentrations (figure 8A). In any case, hourly values did never exceed legal thresholds (200 $\mu g/m^3$ for 1 h and 40 $\mu g/m^3$, for an annual mean concentration of NO_2). Maximum concentration peaks from 100 to 160 $\mu g/m^3$ have been recorded during autumn. The concentration of O_3 is contrary to the NO_x concentration (figure 8B) as it can be expected due to the photochemical reaction among these compounds. O_3 values are also low during the whole year and maximum mean values are recorded in winter and spring with maximum hourly peak concentrations above 100 μg/m^3. (less than 30 times). Legal threshold does not exist for O_3.

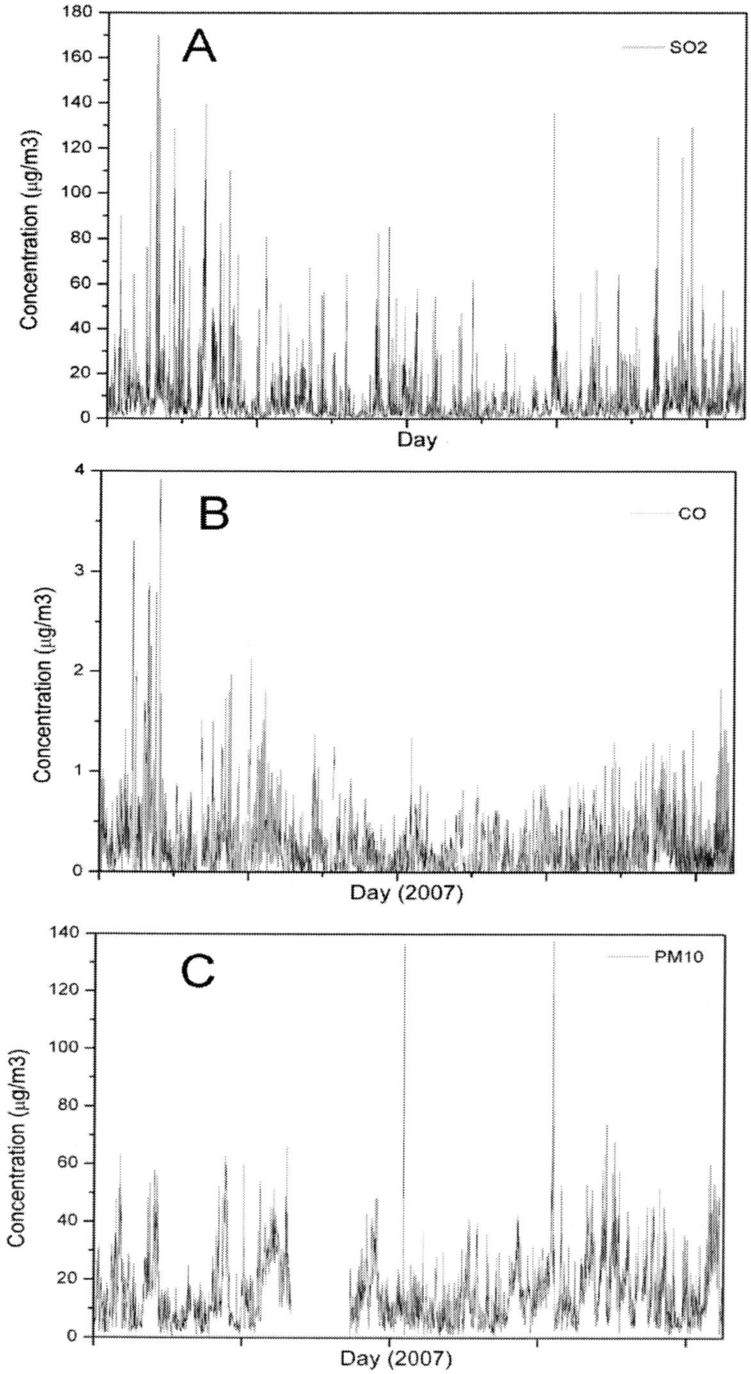

Figure 7. Annual oscillations of gaseous compounds during 2007: (A) SO_2, (B) CO, (C) PM_{10}.

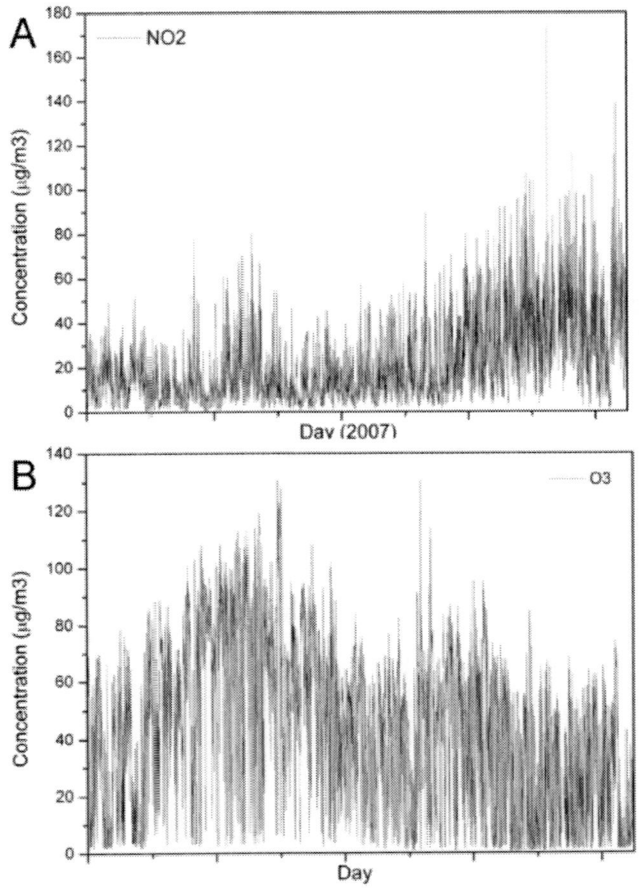

Figure 8. Annual oscillations of gaseous compounds during 2007: (A) NO_2, (B) O_3.

The variability of NO_x and O_3 is related with weather conditions as has been related in the section 1.1. Different chemical processes oxidize NO to NO_2 in the atmosphere, and thus, the most important source of secondary NO_2 is the reaction of NO with O_3 to give NO_3 during the diurnal cycle [23]. Larger days and higher sunlight exposure determine the concentration of NO_x. Rising of O_3 is responsible of faster oxidation of NO to NO_2, enhancing the NO_2 concentration [24]. Moreover, both VOCs and CO are important to determine the O_3 concentration. They have an important effect on the dynamics of NO-NO_2-O_3, as VOCs interrupt this process by being more attractive to the NO that O_3, contributing to O_3 accumulation [24, 29, 30, 31]. In A Coruña city, VOCs are emitted by few industrial activities: the oil refinery and the bioethanol factory. Also, car traffic is an important disperse source (figure 2). The contribution of VOCs to the photochemical formation of O_3 depends on location, season and emission rates [23, 35, 36]. If we consider that emission rates and location are more or less constant in the urban area (as both traffic and industrial activities responsible of VOC emissions are not seasonal activities) the weather seasonal component should be a decisive factor. This asseveration must be checked in further research work.

Particulate matter concentrations are not important in the stations of the city area, most of the time. The legal limit of 2 h mean concentration of 50 $\mu g/m^3$ has been exceeded less than 35 times during 2007 in the city centre. Annual mean concentrations are below the legal threshold of 20 μg/m^3, but values in some suburban stations are next to this value. In fact, two stations have recorded annual mean concentrations above the limit values. Such stations are Lañas and Sorrizo. They are located in a suburban industrial area and a suburban location next to industrial emissions, respectively. Thus, the recorded high values in some urban areas should be related to industrial emissions. Regarding the urban stations (A Coruña-city centre and A Grela), the mean monthly concentrations are highly variable with high standard deviations of the mean ranging from 2 μg/m^3 to 14 μg/m^3 in the city centre (figure 9). If the suburban stations are considered (Arteixo, Lañas, Paiosaco, Sorrizo and Benantes), the standard deviations of the mean are higher and highly variable (from 3 to 23 μg/m^3 and from 4 to 13 μg/m^3). In general, such variability cannot be related to any month or season of the year.

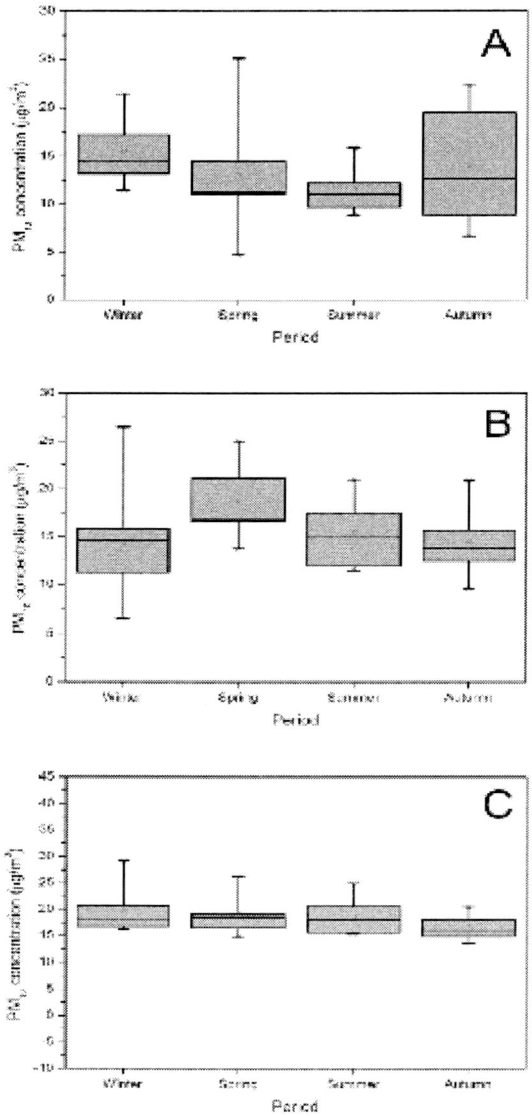

Figure 9. Annual oscillations of particulate matter during 2007 in: (A) urban centre, (B) a suburban industrial location (Arteixo station), and (C) a suburban non-industrial location (Bemantes station).

Conclusion

Industrial and traffic emissions of gaseous and particulate pollutants are evaluated and submitted to air quality directives in the EU. Strategies for air pollution control have been considered in to steps of the emission and diffusion processes or pollutants: establishing a legal register of emissions and measuring air pollution by control stations. Such data have been used and compared in an industrial urban area of the NW of Spain (A Coruña city), to assess the behaviour of pollutants in the area. Inventories of industrial emissions and concentrations measured in control stations of NO_2, NO, NO_x, O_3, SO_2, CO, CO_2 and particulate matter indicate important emission points in the area in 2007. Diffuse emissions due to traffic exhaust must also been considered due to the high number of transport emissions. The hourly, daily, weekly and seasonal behaviour of the pollutants in the area have been studied from data recorded in the control stations.

Both SO_2 and CO showed similar hourly, weekly and seasonal concentrations in all the stations. This is indicative of a same focus, diffusion and behaviour. Similar observations have been done for NO, NO_2, NO_x. As traffic is an important focus of nitrogen oxides, they indicate the important effect of traffic emissions in the concentration of gaseous compounds in air. O_3 showed the contrary behaviour than nitrogen oxides, as it oxidizes NO to NO_2 in the atmosphere during the diurnal cycle. This effect has been observed in different research works. The peaks and increasing trends of O_3 can be related to daily hours, sunlight seasonal duration, and the concentration of VOCs and CO, as they interrupt the oxidation of NO and NO_2.

PM_{10}, $PM_{2.5}$ and PM_1 show a same behaviour in the studied area. The temporal oscillations of particulate matter in air can be related with traffic emissions in the urban centre as the hourly, daily, weekly and seasonal patter of PM fit the pattern of gaseous compounds. Thus, traffic emissions seem to be the main source of air pollutants in the city centre. However, high annual mean PM concentrations have been observed in some suburban areas next to important industrial emission focuses. This is indicative of a different trend: industrial emissions seem to be responsible of PM pollution in suburban areas, where traffic emissions at less abundant.

References

[1] Council Directive 1999/30/EC.

[2] Council Directive 2000/69/EC .

[3] Council Directive 2002/3/EC.

[4] Ministry of Environment of Spain, Registro Estatal de Emisiones y Fuentes Contaminantes, EPER (2010) , http://www.eperes.

[5] Fassina V; Stevan A (1992) Fogwater composition in Venice in relation to stone decay. In: Delgado Rodrigues J, Henriques F, Telmo Jeremias F (ed) Proceedings of the 7th International Congress on Deterioration and Conservation of Stone. *Laboratório Nacional de Engenharia Civil*, Lisbon, 365-374.

[6] Chabas A; Jeannette D (2001) Weathering of marbles and granites in marine environment: petrophysical properties and special role of atmospheric salts. *Environ. Geol* 40, 359-368.

[7] Kumar R; Rani A; Maharaj KK, Srivastava SS (2005) Atmospheric dry deposition to marble and red stone. *Journal of atmospheric chemistry* 50, 243-261.

[8] Englert N. (2004) Fine particles and human health - a review of epidemiological studies. *Toxicology letters* 149, 235-242.

[9] Curtis, L.; Rea, W.; Smith-Willis, P.; Fenyves, E.; Pan Y (2006) Adverse health effects of outdoor air pollutants, *Environment International*, 32, 815-830.

[10] Fellenberg, G. (2000) The chemistry of pollution, Chichester, UK, John Wiley & Sons.

[11] Gaffney, J.S., Marley, N.A. (2009) The impacts of combustion emissions on air quality and climate From coal to biofuels and beyond, *Atmospheric Environment*, 43, 23-36.

[12] Amoroso, G.G.; Fassina, V (1983) Stone decay and conservation: Atmospheric Pollution, Cleaning, Consolidation and Protection, Amsterdam Elsevier.

[13] Fassina V (1988) Environmental pollution in relation to stone decay. Air pollution and conservation. In: Rossvall J, Aleby S (ed) Safeguarding our Architectural Heritage. Elsevier, Amsterdam, pp 133-174.

[14] Laurenzi-Tabasso M; Marabelli M (1992) Il degrado dei Monumenti in Roma in Rapporto all'inquinamento atmosferico. BetaGamma, Viterbo.

[15] Johansson IG; Lindquist O; Mangio RE (1988) Corrosion of calcareous stones in humid air containing SO_2 and NO_2. *Durability Building Materials* 5, 439–449.

[16] Carslaw DC; Beevers SD (2004) Investigating the potential importance of primary NO_2 emissions in a street canyon. *Atmospheric Environment* 38, 3585–3594.

[17] Carslaw DC; Beevers SD, Bell MC (2007) Risks of exceeding the hourly EU limit value for nitrogen dioxide resulting from increased road transport emissions of primary nitrogen dioxide. *Atmospheric Environment* 41, 2073-2082.

[18] Jenkin ME (2004b) Analysis of sources and portioning of oxidant in the UK – Part2: contributions of nitrogen dioxide emissions and background ozone at a kerbside location in London. *Atmospheric Environment* 38, 5131-5138.

[19] Roorda-Knape MC; Janssen NAH; de-Hartog J; van-Vliet PHN; Harssena H; Brunekreef B (1999) Traffic related air pollution in city districts near motorways. *The Science of Total Environment* 235, 339-341.

[20] Gilbert NL; Woodhouse S; Stieb DM; Brook JR (2003) Ambient nitrogen dioxide and distance from a major highway. *The Science of Total Environment* 312, 43-46.

[21] European Environment Agency (2007) Air pollution in Europe 1990-2004. EEA Report 2/2007. ISSN 1725-9177.

[22] Pleijel H; Klingberg J; Bäck E (2009) Characteristics of NO_2 pollution in the city of Gothenburg, South-West Sweden – Relation to NO_x and O_3 levels, photochemistry and monitoring location. Water, Air and Soil Pollution: Focus 9, 15-25.

[23] Jenkin ME; Clemitshaw KC (2000) Ozone and other secondary photochemical pollutants: chemical processes governing their formation in the planetary boundary layer. *Atmospheric Environment* 34, 2499-2527.

[24] Jenkin ME; Utembe SR, Derwent RG (2008) Modelling the impact of elevated primary NO_2 and HONO emissions on regional scale oxidant formation in the UK. *Atmospheric Environment* 42, 323-336.

[25] Martin P; Cabañas B; Villanueva F; Gallego MP; Colmenar I; Salgado S (2010) Ozone and Nitrogen Dioxide Levels Monitored in an Urban Area (Ciudad Real) in central-southern Spain. Water, *Air and Soil Pollution* 208, 305-316

[26] WHO (2000) Air quality guidelines for Europe. European Series, No 91 (2nd ed) Copenhagen, Denmark: WHO Regional Publications.

[27] Parmar GR; Rao NN (2009) Emerging Control Technologies for Volatile Organic Compounds. Critical Reviews in Environmental Science and Technology 39, 41–78.

[28] Virgarzan R (2004) A review of sulfate ozone background levels and trends. *Atmospheric Environment* 38, 3431-3442.

[29] Haagen-Smit AJ; Fox MM (1954) Photochemical ozone formation with hydrocarbons and automobile exhaust. *Journal of Air Pollution Control Association* 4, 105-109.

[30] Haagen-Smit AJ; Fox MM (1956) Ozone formation in photochemical oxidation of organic substances. *Indian Enginering Chemistry* 48, 1484.

[31] Leighton PA (1961) Photochemistry of Air Pollution. Academic Press, New York.

[32] Sillman S; Samson FJ (1995) Impact of temperature on oxidant photochemistry in urban, polluted rural and remote environments. *Journal of Geophysical Research* 100, 11497-11508.

[33] Rudd HJ (1995) Emissions of volatile organic compounds from stationary sources in the United Kingdom: speciation. AEA Technology Report, AEA/CS/REMA-029. AEA Technology. Oxfordshire, UK.

[34] Guenther A; Hewitt CN; Erickson D; Fall R; Geron C; Graedel T; Harley P; Klinger L; Lerdau M; McKay WA; Pierce T; Scholes B; Steinbrecher R; Tallamraju R; Taylor J; Zimmerman P (1995) A global model of natural volatile organic compound emissions. *Journal of Geophysical Research* 100, 8873-8892.

[35] Sillman S (1999) The relationship between ozone, NOx and hydrocarbons in urban and polluted rural environments. *Atmospheric Environment* 33, 1821-1845.

[36] Kleinman LI (1991) Seasonal dependence of boundary layer peroxide concentration: the low and high NOx regimes. *Journal of Geophysical Research* 96, 20721-20733.

[37] Jenkin ME (2004a) Analysis of sources and portioning of oxidant in the UK - Part 1: the NOx-dependence of annual mean concentrations of nitrogen dioxide and ozone. *Atmospheric Environment* 38, 5117-5129.

[38] Itano Y; Bandow H; Takenaka N; Saitoh Y; Asayama A; Fukuyama J (2007) Impact of NO_x reduction on long-term ozone trends in an urban atmosphere. *Science of Total Environment* 379, 46-55.

[39] Global Carbon Project (2010) [http://www.globalcarbonproject.org /carbonbudget/08/hl-full.htm]

[40] Pueschel RF; VanValin CC; Castillo RC; Kandlech RC; Ganor E (1986) Aerosols in polluted versus non polluted air masses: long range transport and effect on clouds, *Journal of Climate and Applied Meteorology*, 25, 1908-1917.

[41] Janhäll S; Olofsson KFG; Anderson PU; Pettersson JBC; Hallquist M (2009) Evolution of the urban aerosol during winter temperature inversion episodes, *Atmospheric Environment*, 40, 5355-5366.

[42] Vardoulakis S; Kassomenos P (2008) Sources and factors affecting PM10 levels in two European cities: Implications for local air quality Management, *Atmospheric Environment*, 42, 3949-3963.

[43] Spengler J; Brauer M; Koutrakis P (1990) Acid air and health. *Environmental Science and Technology* 24, 946-956.
[44] Chow JC (1995) Measurement methods to determine compliance with ambient air quality standards for suspended particles, *Air and Waste Management Association*, 45, 320-382.
[45] Chow JC; Watson JG; Green MC; Lowenthal DH; Bates B; Oslund W; Torres G (2000) Grossborder transport and spatial variability of suspended particles in Mexicali and California's Imperial Valley, *Atmospheric Environment*, 34, 1833-1843.
[46] Chan YC; Simpson RW; Mctainsh GH; Vowles PD; Cohen DD; Bailey GM (1997) Characterisation of chemical species in PM2.5 and PM10 aerosols in Brisbane, Australia, *Atmospheric Environment*, 31, 3773–3785.
[47] Chan YC; Simpson RW; Mctainsh GH; Vowles PD; Cohen DD; Bailey GM (1999) Source apportionment of PM2.5 and PM10 aerosols in Brisbane (Australia) by receptor modelling, *Atmospheric Environment*, 33, 3251-3268.
[48] Harrison RM; Deacon AR; Jones MR; Appleby RS (1997) Sources and processes affecting concentrations of PM10 and PM2.5 particulate matter in Birmingham (U.K.), *Atmospheric Environment*, 31, 4103-4117.
[49] Hildemann LM; Markowski GR; Cass G. (1991) Chemical composition of emissions from urban sources of fine organic aerosol, *Environmental Science and Technology*, 25, 774-759.
[50] Van Grieken R; Torfs K (1996) Atmospheric aerosols and deposition near historic buildings: chemistry, sources, interrelations and relevance. In: Origin, mechanisms and effects of salts on degradation of monuments in marine and coastal environments, Zezza F., ed., Bary, Italy, *European Comission Research Workshop*, 23-35.
[51] Pio CA: Ramos MA; Duarte AC (1998) Atmospheric aerosol and soiling of external surfaces in an urban environment, *Atmospheric Environment*, 32, 1979-1989.
[52] Song XH; Polissar AV; Hopke PK.(2001) Sources of the particle composition in the northeastern US, *Atmospheric Environment*, 35, 5277-5286.
[53] Singh H; Kumar PK (2009) Analysis of coal ash for race elements and their geoenvironmental implications. *Water Air Soil and Pollution* 198, 87-94.
[54] Krupa SV (2002) Sampling and physico-chemical analysis of precipitation: a review. *Environmental Pollution* 120:565-594..
[55] Maas S; Scheifler R; Benslama M; Crini N; Lucot E; Brahmia Z; Benyacoub S; Giraudoux P. (2010) Spatial distribution of heavy metal concentrations in urban, suburban and agricultural soils in a Mediterranean city of Algeria, *Enviromental Pollution*, 158, 2294-2301.
[56] Wu Y; Hao J; Fu L; Wang Z; Tang U (2002) Vertical and horizontal profiles of airborne particulate matter near major roads in Macao, *China Atmospheric Environment*, 36, 4907-4918.
[57] Bityukova L; Shogenova A; Birke M (2000) Urban geochemistry: a study of elements distribution in the topsoil's of Tallinn (Estonia). *Environmental Geocheistry and Health* 22, 173-195.
[58] Chabas A; Lefèvre R (2000) Chemistry and microscopy of atmospheric particulate at Delos (Cyclades, Greece) *Atmospheric Environment* 34, 225-238.
[59] Pierson WR; Brachaczek WW (1983) Particulate matter associated with vehicles on the road II. *Aerosol Sci. Tech.* 2, 1-40.

[60] Carballeira A; Devesa C; Retuerto R; Santillán E; Ucieda F (1983) Bioclimatología de Galicia. A Coruña, Spain, Fundación Pedro Barrié de la Maza Conde de Fenosa.

[61] Ministry of Development of Spain, Puertos del Estado (2010) http://www.puertos.es/es/oceanografia_y_meteorologia/banco_de_datos/viento.html,

[62] Xunta de Galicia, Calidade do aire (2010) http://medioambiente.xunta.es/calidadeAire/docsInterese.jsp.

[63] Sanjurjo-Sánchez J; Vidal-Romani JR; Alves CAS, Fernández-Mosquera D (2009) Origin of gypsum-rich coatings on Historic Buildings. *Water, Air and Soil Pollution* 204, 53-68.

[64 Sanjurjo-Sánchez J; Vidal-Romaní JR; Alves CAS (2011) Deposition of particles on gypsum-rich coatings of Heritage buildings in urban and rural environments. *Construction and Building Materials* 25, 813-822.

In: Air Pollution Emissions
Editors: Daniel G. Vasiliev and Robert A. Kask
ISBN: 978-1-62100-453-0

Chapter 5

VOLATILE ORGANIC COMPOUNDS (VOCS) IN THE ATMOSPHERE. AN OVERVIEW OF THE SAMPLING AND ANALYSIS TECHNIQUES

Débora Pérez-Rial*[*1], *Purificación López-Mahía*[1,2], *Soledad Muniategui-Lorenzo*[1] and *Darío Prada-Rodríguez*[1,2]

[1]Department of Analytical Chemistry, University of A Coruña, A Coruña, Spain
[2]University Institute of the Environment, University of A Coruña, A Coruña, Spain

ABSTRACT

Volatile organic compounds (VOCs) constitute a heterogeneous group of substances with different physical and chemical behaviours but with the common feature of having high vapour pressures. When these compounds reach the atmosphere they undergo fast reactions with ozone, OH and nitrate radicals and, as a consequence of these reactions, it is produced a great variety of pollutants. These secondary pollutants include ground-level ozone, peroxyacetyl nitrate, carbonyl compounds or organic acids. Additionally, the generated products also contribute to the formation of secondary organic aerosols (SOA) in the atmosphere. Apart from these harmful effects on the environment, VOCs also produce acute and chronic effects on human health that may range from simple sensory irritations to severe behavioural, neurotoxic and genotoxic damages.

In the monitoring programs the most suitable analytical method has to be selected taking into account the nature of the analytes of interest, the composition of the matrix and the concentrations of the target analytes in such matrix. The assurance of a representative sampling, the correct manipulation and storage of the sample and the precise and accurate determination of the target analytes are key parameters in the environmental monitoring programmes.

The aim of this chapter is to give an overview of the methodologies used for the sampling and analysis of VOCs in the atmosphere summarizing their main advantages and limitations. The methods discussed here include the traditional whole air sampling

* Corresponding author: dperez@udc.es (D. Pérez-Rial) Tel.:+ 34 981167000 Extension 2187, Department of Analytical Chemistry, University of A Coruña, Campus da Zapateira s/n, E-15071, A Coruña, Spain.

techniques, the cryogenic trapping and the adsorptive enrichment on solid adsorbents in the sampling stage and the use of gas chromatography combined with different detection systems for the separation and detection of the target analytes. The possible applications of several newer analytical methods for the analysis of VOCs in air are also discussed.

1. Introduction

Volatile Organic Compounds (VOCs) are a group of substances characterized by their high vapour pressures. There is not an unambiguous definition for this category of compounds and it is frequent that they are divided into different subclasses as, for example, Non Methane HydroCarbons (NMHC) which is strictly referred to hydrocarbons containing only hydrogen and carbon atoms different from methane, Oxygenated VOCs (OVOCs) which specifically includes compounds containing oxygen atoms or Biogenic VOCs (BVOCs) which is referred to the subclass of VOCs that are emitted from natural sources.

The presence of such substances in the atmosphere is a direct consequence of their release from both anthropogenic and biogenic sources. Usually, the anthropogenic sources are more localized than the biogenic ones and they can be divided into stationary and mobile sources. In the year 1989 Cullis and Hirschler [1] estimated that the human activity was responsible for the emission of 235 millions of metric tonnes of VOCs per year into the atmosphere, since Guenther and cols. [2] fixed the NMHC emissions in 1150 TgC yr^{-1}, being this value more than a factor of 7 greater than estimated global anthropogenic emissions.

The presence of VOCs in the atmosphere with an anthropogenic origin is due to sources as combustion processes, especially those related to the combustion of fossil fuels [3-8], petrochemical activities and petroleum refinery processes [9-11], the use of solvents with both domestic and industrial purposes [12, 13] or road traffic [14-17]. Apart from these sources, evaporative emissions also constitute an important input of VOCs in the atmosphere. This kind of emissions includes vehicular evaporative emissions, evaporative losses related to petrol loading and unloading activities, surface coating activities, dry cleaning processes, graphical art applications, etc [18].

The characteristics and magnitude of the emitted VOCs will depend on the kind of source, for example alkylbenzenes and light n-alkanes are released by the road traffic while heavier n-alkanes are produced by vehicles using diesel fuel [19]. The propane can be emitted in processes to obtain energy, in biofuel and biomass burning and in different industrial activities [20]. On the other hand, some chlorinated compounds as tetrachloroethylene are released in dry-clean processes while other substances as trihalomethanes are generated in waterworks plants, reaching the atmosphere by evaporation. 1,1,1-Trichloroethane (also called methyl chloroform) has been used as a solvent in the industry, although Rudolph and cols. [21] have also reported its emission in biomass burning processes. Toluene, ethylbenzene, xylene isomers, methylethylketone, ethylenglycol, n-hexane and acetone have a common origin in the paints used in the naval industry [22] and, at the same time, the BTEX fraction formed by Benzene, Toluene, Ethylbenzene and Xylenes suffers an increment in its emissions as a consequence of its production by road traffic sources.

Biogenic emissions of VOCs are mainly dominated by their release from the terrestrial vegetation but there are also other minor sources as the oceans, the continental surface waters, the soils, the sediments, the geological deposits of fossil combustibles, the microbial decomposition of the organic material and the volcanic emissions [2].

The presence of this kind of volatile substances in the atmosphere produces severe detrimental effects on the environment and also on human health. VOCs may react to a high extent with atmospheric constituents such as ozone or hydroxyl and nitrate radicals producing a great variety of oxidized compounds [23, 24]. As a consequence of such reactions, these compounds participate in the chemical production of oxidants, greenhouse gases and acid substances and, obviously, they also take part in the global budget of carbon in the atmosphere. Besides this, VOCs also contribute to the formation of secondary organic aerosols [25-31] by means of their condensation on previously formed particles or acting as condensation nuclei [32] and these aerosols can have a direct and an indirect influence on the radioactive budget of the Earth [33].

As a consequence of the previously cited harmful effects on both the environment and human health, a growing concern to reduce the VOC levels in the atmosphere has emerged. This interest has been reflected in the abatement policies adopted by many countries but also in the increasing number of publications devoted to the determination and characterization of the concentrations and sources of VOCs in the air and to the modeling of their emissions.

One of the main problems related to the analysis of VOCs in the air is that the atmospheric concentrations of these compounds are at levels of parts per billion in volume or even lower and, besides this, the complexity of the environmental air matrix makes it difficult the reliable determination of these substances. For these reasons the combination of preconcentration sampling devices with efficient separation methods and sensitive detection systems for the analysis of VOCs in the air is required. In this context, the establishment of reference calibration methods and quality control systems gains importance to obtain reliable results [34].

In this chapter it is given an overview of the traditional methods used for the sampling and analysis of VOCs in the atmosphere. These methods include the whole air sampling techniques and the adsorptive enrichment on solid adsorbents and also the use of the conventional chromatographic techniques combined with different detection systems for the determination and eventual quantification of the compounds. Additionally other techniques, with a more recent development, as solid phase microextraction for the VOC sampling or comprehensive two-dimensional chromatography for the separation of these complex mixtures are also discussed.

2. Sampling Methods

One of the first problems that emerges in the analysis of environmental matrices is the sampling and manipulation of the sample. It is frequent that the samples are too much diluted or that they are very complex to be directly compatible with the techniques used for the analytical determination. For these two reasons samples need to be subjected to several previous procedures in order to ensure their compatibility with the analytical instrumentation guaranteeing, at the same time, an effective separation and identification and an adequate quantification of the individual analytes.

Whereas in the analysis of soil and aqueous matrices the sampling and the preconcentration steps are usually clearly differenced stages, in the work with gaseous samples these two procedures can be carried out simultaneously.

Air is a very complex matrix to develop a correct sampling because it constitutes a heterogeneous system formed by gases, liquids (aerosols) and particles with a variable and changing composition. In that system the pollutants are subjected to several processes as diffusion, dispersion and chemical reactions which produce secondary pollutants that are usually more toxic than the original ones.

Air sampling shows several difficulties that the analyst has to face to. First, the atmospheric pollutants are frequently at very low concentrations, generally in the range of a few parts per billion or even lower and, for this reason, the standards prepared to validate the method should contain the analytes at such low levels, making it difficult their manipulation and preparation. Besides this, it is very important to handle the samples with care enough to prevent the introduction of contamination that alters the original composition of the samples and avoiding, at the same time, losses of those compounds with the highest volatilities [35, 36].

Sampling methods have to be as simple as possible but simultaneously they have to guarantee that the taken aliquot is representative of the air mass that is being subjected to the sampling, without disturbances from a qualitative or a quantitative point of view. It is also important to keep under control the absence of losses, degradation of the analytes or contamination of the sample in the sampling stage or during the time previous to the analysis, including the transport and storing stages [37, 38].

As a consequence of the low levels of VOCs in ambient air, it is usual that the analytical methods include a preliminary stage to achieve the enrichment of the samples with the target analytes in order to improve the sensitivity of the methods and to reach appropriate detection and quantification limits. However, the preconcentration of the samples must not be excessive because it could produce the saturation of the systems used for the separation and detection of the analytes or, even when this saturation was not effective, the response produced by the instrumental equipment could be outside of the response linear range.

The methodologies used for the analysis of VOCs in atmospheric samples can be divided into on-line methods, operating in real time, and *ex-situ* methods [39]. The on-line methods, allow the caption of a volume of air in a continuous way and they integrate a detector system in the same instrument, providing a signal practically in real-time. The *ex-situ* methods, or off-line methods, are based on the trapping of the analytes using an appropriate support or on the collection of a volume of air to carry out an ulterior analysis in the laboratory.

The techniques for the on-line sampling, in real or quasi-real time, have a great number of advantages in comparison with the indirect methods, especially because of the absence of a preconcentration stage in most of the cases. However, *ex-situ* methods have been traditionally more used than the on-line ones.

The traditional systems used for the sampling of VOCs in ambient air include the whole air (isolation) sampling techniques, the cryogenic trapping and the retention of the analytes onto adsorbent solids. These last two methods show the advantage of allowing the sampling and the preconcentration of the analytes simultaneously, whereas in the sampling of a whole volume of air this enrichment of the samples with the target analytes is not produced in the sampling stage. For that reason, this technique is often combined with an ulterior preconcentration stage, previous to the chromatographic separation.

2.1. Whole Air Sampling

Whole air sampling involves the caption of a volume of air in a receptacle with adequate characteristics which is transported to the laboratory to its subsequent preconcentration and analysis. Traditionally, this method has been used for the sampling of compounds with low polarities and nowadays it finds its major application in the sampling of very light compounds (C_2-C_4) that cannot be retained on adsorbent solids without using a cryogenic system in the sampling stage.

The containers used for the sampling of VOCs have to be chemically inert and they should guarantee the integrity of sample, minimizing the leaks, the diffusion and the permeation of the analytes through their walls [40]. Generically, there are two kinds of containers: the rigid ones, as for example the metallic canisters, and the flexible containers, typically in the form of sampling bags made of a polymeric film with a suitable composition. In the category of rigid containers the glass balloons are also included, however both the limitation of the sampled volume and their fragility have restricted the use of such devices for the VOC sampling.

Table 1 compiles several references in which the sampling stage was carried out by means of the isolation of a whole air volume using a proper receptacle.

The most important advantages related to the sampling capturing a whole air volume are:

- The simplicity of the systems.
- The possibility of analysing several aliquots of a same sample.
- The absence of discrimination among the sampled compounds. That makes these systems suitable for the analysis of samples with an unknown composition.

The main inconveniences of the methods capturing a whole air sampling are:

- The limitation of the sampling volume to a few litres. This limitation could make this kind of methods not suitable for the analysis of VOCs at very low concentrations.
- The absence of stability of the analytes for a long time in the sampling devices because of processes of adsorption on the walls of the containers, permeation, interactions among the different components of the sample, reaction of the target analytes with compounds from the wall of the container, etc.
- Additionally, the use of techniques for the whole air sampling involves the incorporation of a preconcentration step to enrich the samples with the target analytes before their analysis. Adsorbent solids and cryogenic trapping are the methods more commonly used for this purpose [41].

2.1.1. Inert Bags

Inert bags are containers made of flexible plastic materials with variable compositions. The sampling is carried out forcing the entrance of air into the bag by means of a pump connected to a sampling port.

Table 1. Examples of whole air sampling for VOC analysis

Sampling(1)	Preconcentration(2)	Ulterior analysis(3)	Application	Citation (year)
Opaque Tedlar® bags	Tenax GC® cooled by N_2 (l)	GC/FID	Characterization of vehicular emissions	[42] (1990)
Glass bottles	Glass beads cryogenically cooled	GC/MSD	Determination of volatile halocarbons in volcanic emissions	[43] (2000)
Bags	A polymeric adsorbent (GDX-102®) and ulterior cryoconcentration with N_2(l)	GC/MSD; GC/FID	Study of the temporal evolution of VOC concentrations in Changchun (China)	[44] (2000)
Tedlar® bags	Carbotrap® + Carbosieve SIII® and cryogenic trapping	GC/MSD	Measurement of VOC emitted by vehicles powered with different combustibles	[45] (2000)
Canister	Cryogenic trapping on glass beads	GC/MSD	Determination of toxic compounds in the air of Minnesota (EE.UU.)	[46] (2000)
Canister	Cryogenic trapping on glass beads using N_2(l)	GC/MSD; GC/FID; GC/ECD	Measurements of VOCs over the Pacific Ocean taking samples with an aeroplane	[47] (2001)
Canister	Cryogenic trapping with N_2(l) and ulterior trapping in glass beads cooled with N_2(l)	GC/FID; GC/MSD;	Study of the VOC seasonal variations in Seoul (Korea)	[48] (2001)
Canister	Cryogenic concentrator	GC/MSD	Determination of VOCs in indoor air sampled in houses of Hong-Kong	[49] (2002)
Canister (passive sampling)	Cryoconcentration on glass beads	GC/FID	Identification of the VOC (C_2-C_{17}) sources in Helsinki, using chemical mass balance and multivariate receptor models	[50] (2003)
Canister	Cryogenic trapping	GC/MSD	Determination of VOCs in the ambient air of Taiwan	[51] (2003)
Canister	Cryogenic trapping at -150°C	GC/MSD	Determination of VOCs released from different surfaces covered with varnish	[52] (2003)
Tedlar® bag/Canister	Cryogenic trapping	GC/FID	Determination of VOCs released in the combustion of different fuels in domestic cookstoves in China	[53] (2003)
Canister	Cryogenic trapping in two stages	GC/FID	Determination of the photochemical age of air masses in an urban area of Phoenix (EE.UU.)	[54] (2003)
Canister	Cryogenic trapping	GC/MSD	Measurement of the VOC levels in the public transport of Hong Kong	[55] (2003)
Canister	Cryogenic trapping	GC/MSD	Measurement of the VOCs emitted by varnish layers of different thickness	[56] (2003)
Canister	N_2(l) cryogenic trap and ulterior trapping on glass beads cryogenically cooled by N_2(l)	TD/GC/FID TD/GC/MSD	Determination of VOCs in the atmosphere of Seoul (Korea) during the daytime	[57] (2003)
Nalophan® and Calibond® bags	Not used	Olfactometry	Study of the stability of odours in different kinds of sampling bags	[40] (2003)

Sampling[(1)]	Preconcentration[(2)]	Ulterior analysis[(3)]	Application	Citation (year)
Tedlar® bags	Not specified	GC/FID GC/ECD	Determination of VOC emissions during cold-starts of motorcycles	[58] (2003)
Rigid containers	Cryogenic trapping	GC/MSD	Measurement of trihalomethanes in air in the surroundings of Cape Grim (Tasmania)	[59] (2004)
Canister	Tenax GC® + Carboxen 1000®	TD/GC/FID	Determination of the exposure levels to benzene in the public transport of Mexico DF	[60] (2004)
Canister (passive integrated sampling)	Cryogenic trapping	GC/MSD	Evaluation of the risk of developing a cancer as a consequence of the exposition to VOCs in different environments	[61] (2004)
Canister	Cryogenic trapping	GC/MSD	Determination of the NMHC sources in Hong Kong using receptor models	[62] (2004)
Canister	Preconcentrator in three stages	GC/FID	Study of the transport of several pollutants along the Chinese coast to other Western Pacific areas	[63] (2004)
Canister	Cryogenic trapping on glass beads	GC/MSD	Determination of VOCs in urban atmospheres of Minneapolis (EE.UU.). Comparison with the results obtained using personal organic vapor monitor system	[64] (2004)
Canister	Not specified	GC/MSD	Characterization of the VOC released during the combustion of incense	[65] (2004)
Canister	Not specified	Bidimensional GC/FID	Measurement of the emissions produced by vehicles in the interior of a tunnel in Houston (EE.UU.)	[66] (2004)
Canister	Cryogenic trapping (Method TO-15 [67])	GC/MSD	Determination of benzene in several locations of New York (EE.UU.) among 1990-2003. Comparison of canister and adsorbent methodologies	[68] (2005)
Canister	Not specified	Photoacoustic spectrometry	Determination of ethene in atmospheric samples taken in Mexico City	[69] (2005)
Canister (passive sampling)	Cryogenic trapping with $N_2(l)$	GC/FID GC/ECD GC/MSD	Determination of methane and NMHCs in 45 Chinese cities	[70] (2005)
Tedlar® bag	Tenax TA®: Carbosieve SIII® and ulterior cryoconcentration	GC/FID	Study of the seasonal variations in monoterpene emissions produced by coniferous trees in Korea. The sampling was developed using a branch enclosure system	[71] (2005)

Table 1. (Continued)

Sampling(1)	Preconcentration(2)	Ulterior analysis(3)	Application	Citation (year)
Tedlar® bag	Multilayer adsorbent trap (porous polymer: graphitized carbon black: two carbon molecular sieves)	TD/Bidimensional GC/FID	Measurement of VOC emissions produced by vehicles. Study of the VOC levels found in Bresso, in the outskirts of Milan using chemical mass balance and positive matrix factorization	[72] (2005)
Canister	Cryogenic trapping	GC/FID GC/ECD GC/MSD	Determination of the hydrocarbons emitted by a commercial airliner	[73] (2006)
Canister	Cryogenic trapping	GC/MSD or GC/FID	Determination of the VOC fraction released from a grassland in Mongolia	[74] (2006)
Canister	Cryogenic trapping with N_2(l)	GC/FID GC/ECD GC/MSD	Determination of volatile halocarbons in 45 Chinese cities	[75] (2006)
Canister (passive sampling)	Cryoconcentration with N_2(l) on a bed of glass beads	GC/MSD GC/FID GC/ECD	Determination of methane, NMHC and CO in different cities from the EE.UU.	[76] (2008)
Canister (passive sampling)	Cryoconcentration with N_2(l) on glass beads	GC/MSD GC/FID GC/ECD	Determination of VOCs in two urban areas of Pearl River Delta (China)	[77] (2008)
Tedlar® bags	Cold trap of Carbopack B® and Carbopack C®	GC/FID	Comparison between chromatographic methods and olfactometric techniques	[78] (2008)
Canister	Cryogenic trapping	GC/FID	Determination of the VOC emissions produced by vehicles powered by diesel fuels	[79] (2008)
Canister	Cryogenic concentration	GC/MSD	Evaluation of the exposition to VOCs in interior air sampled in houses of Ontario (Canada)	[80] (2008)
Canister	Cryoconcentration with N_2(l) on a bed of glass beads	GC/MSD GC/FID GC/ECD	Determination of VOCs in urban and rural air from Pearl River Delta (China)	[81] (2008)
Canister	Preconcentrator for VOCs in three stages	GC/MSD	Characterization of the emission profiles of VOCs produced by different sources in Shanghai (China). Application of a receptor model to identify the individual contribution of each identified source	[82] (2009)
Canister	Cryoconcentration with N_2(l) on a bed of glass beads	GC/FID GC/ECD GC/MSD	Study of the gaseous emissions coming from a soil covered with snow and from the accumulated snow on it	[83] (2009)
Transparent and opaque Tedlar® bags, Nalophan®, Teflon® and FlexFoil® bags	SPME (fibres of polydimethylsiloxane and Carbowax®)	TD/GC/MSD	Study of the stability of sulphured VOCs contained in different kinds of polymeric bags	[84] (2009)

Sampling(1)	Preconcentration(2)	Ulterior analysis(3)	Application	Citation (year)
Tedlar® bags	SPME and different adsorbents: Tenax TA®, Chromosorb 106®, Carbotrap X®, Carbopack X®, Carboxen 509®, Unicarb®, Carboxen 1000®, Carboxen 1003® (used both individually and combined)	TD/GC/MSD	Optimization of a preconcentration method for sulphured VOCs emitted during the breath exhalation	[85] (2009)
Canister	A bed of glass beads cooled with N_2(l)	GC/FID GC/ECD GC/MSD	Monitoring of the levels of CO, NMHC, halogenated VOCs, C_1-C_4 alkylated nitrates and sulphured compounds in the South Pole from the year 2000 to the year 2006	[86] (2010)
Canister	Cryoconcentration	GC/MSD GC/FID	Determination of the fluxes of non- methane VOCs emitted by six dairies in California (EE.UU.)	[87] (2010)
Canister	Preconcentrator for VOCs in three stages	GC/FID	Measurement of the emission of VOCs in six different locations of Kanto (Tokyo)	[88] (2010)
Canister	A bed of glass beads cooled with N_2(l)	GC/FID; GC/MSD; GC/ECD	Determination of VOCs in a suburban area of Mexico. Comparison with an on-line method	[89] (2011)
Canister (integrated sampling)	A bed of glass beads cooled with N_2(l)	GC/FID; GC/MSD; GC/ECD	Study of the effect of the driving speed and fuel on the emissions of NMHC produced by cars	[90] (2011)
Canister	Cryoconcetration	GC/FID	Validation of a remote sensing methodology for the estimation of isoprene emitted by vegetation in Honk Kong	[91] (2011)
Tedlar® bags	SPME	GC/MSD	Determination of light sulphuric compounds, fatty acids, indolic and phenolic compounds in a swine nursery house in Seoul	[92] (2011)

(1) Unless otherwise indicated, the sampling stages were developed forcing the entrance of air towards the sampling device (active sampling).

(2) N_2(l): liquid nitrogen SPME: Solid Phase MicroExtraction.

(3) TD: Thermal Desorption.

GC: Gas Chromatography FID: Flame Ionization Detector MSD: Mass Spectrometry Detector ECD: Electron Capture Detector.

The use of inert bags is quite frequent to sample compounds with a high pressure of vapour and also when the concentrations of the target analytes are high. These devices are frequently used for the study of odour episodes by means of olfactometric techniques.

Nowadays, there is commercially available a wide variety of sampling bags made of different materials. One of the substances more frequently used for this purpose is the PolyVinyl Fluoride (PVF) and the bags made of this polymer are commercialized under the denomination of Tedlar® bags (Dupont de Nemours). The bags made of copolymers of Fluorinated Ethylene and Propylene (FEP) are commercialized as Teflon® bags (Dupont de Nemours) and those made of PoliEthylene Terephtalate (PET) copolymer are called Nalophan® bags (Kalle Nalo GmbH) [40, 93]. Apart from these models, there are also commercially available different trade marks with similar benefits as the Halar®, Saran® or Mylar® bags.

The advantages of the use of inert bags for VOC sampling are the easiness of their handling, their low cost, the wide range of compositions and thickness of the films available in the market and the possibility of reuse them after a simple cleaning programme. This cleaning stage consists of the filling of the bags several times with zero air, with an ultra-high purity, and of the ulterior evacuation of the air using a pump. The cleaning programme has to be very exhaustive to avoid memory effects and besides this, it is recommended not to fill the bags over the 80% of their maximum volume to reduce the risk of explosion.

One of the main inconveniences reported in the literature for the VOC sampling using polymeric bags is that there can be losses of the sampled air. These losses may be caused by imperfections in the bag surface, chemical reaction of the compounds retained in the bag either with the bag walls or with other compounds contained in the same volume of air, the adsorption of the target analytes on the polymeric films and the permeation of the analytes through the walls of the bags.

To avoid these losses by adsorption or permeation processes multilayer bags have been developed. Thus, for example, Cali-5-bond® bags are formed by a structure composed by a layer of 75 μm of polyethylene, a layer of 40 μm of polyamide, an aluminium layer of 12 μm, a film of 3-4 μm of polyvinyl chloride and a film of 12 μm of polyester. This multilayer configuration provides more resistance and minimizes the permeation processes. Other disadvantage of the use of bags is that, although the flexibility of the walls makes easier the transport and storage, it is also responsible for the high risk of explosion and drilling.

In a study of the stability of VOCs in Tedlar® bags, Stump and Dropkin [94] found that the samples could be stored in this kind of containers for a period of up to six hours without significant losses of the analytes, getting quantitative recoveries for all of them. Cariou and Guillot [93] extended the stability period obtained with double-film Tedlar® bags for aromatic compounds up to 46 hours, although these authors reported significant losses in the halogenated VOCs during the same period of time. In the study of the odour episodes in a tobacco processing factory by olfactometric methods, Van Harreveld [40] found that the Nalophan® bags kept the odours stable during 12 hours, and these odours were reduced to the half after a period of 30 hours. In the same study, the stability of the odour reported in Cali-Bond® metallic bags was lower.

Nowadays the use of polymeric bags has great applications in the study of BVOC emissions produced by vegetation using enclosure methods. In these methodologies a small part of a plant (a branch, a twig, a leaf, etc) or the whole plant is enclosed in a receptacle, or cuvette, especially designed for that purpose. At the same time, ambient air is pumped into the enclosure system and, after a period of stabilization, the VOC emissions are monitored in the outgoing air current exiting from the enclosure system. These cuvettes are made of materials transparent to the solar radiation and, for that reason, the use of plastic bags has a wide application in this field. For example Boissard and cols. [95] used Teflon® bags as enclosure systems to carry out the study of the seasonal emission cycles produced by the species *Ulex europaeus* in the United Kingdom. Harrison and cols. [96] also utilized an enclosure system using bags made of the same kind of material for the determination of the isoprene and monoterpene emissions produced by *Abies Borisii-regis* in Greece. Sabillón and Cremades [97] studied the diurnal and seasonal variations of the terpenoid emissions produced by *Pinus pinea* and *Quercus ilex* in a Spanish Mediterranean area using Teflon® bags as an enclosure system. Simon and cols. [98] determined the emissions produced by the species *Pinus halepensis* using an enclosure system also made of Teflon® while Lin and cols. [99] determined the monoterpene emissions produced by the roots of *Pinus pinea* species using a Teflon® bag as the enclosure system. Apart from Teflon® bags, Tedlar® [100, 101] and polyethylene and PET bags are used as enclosure systems for the characterization of BVOCs produced by vegetation [102-104].

2.1.2. Canisters

Canisters are containers with rigid walls made of aluminium or stainless steel. The use of these devices for the sampling of VOCs in air was adopted by the United States Environmental Protection Agency (US EPA) in the eighties with the publication of the methods EPA TO-14 [105] and TO-14A [106] for the analysis of 40 non-polar compounds. In the year 1999 these methods were revised and, as a consequence of this, it was published a new method (TO-15 [67]) for the analysis of 97 VOCs, including polar compounds, in which the sampling was also based on the use of canisters.

Nowadays, the sampling by canisters has been applied to the analysis of more than 150 polar and non-polar VOCs in ambient air, with concentration levels ranging from parts per billion in volume (ppbv) to parts per million in volume (ppmv). This methodology has been validated for the determination of some VOCs in the atmosphere at a level of parts per trillion (pptv) [41].

The stability of the VOC contents in samples taken with canisters made of untreated steel is not guaranteed. For that reason the canister used in the sampling of VOCs are manufactured using stainless steel with Grades 304 or 316. The Grade 316 stainless steel is more resistant to corrosion than the Grade 304 stainless steel.

In order to passivate the walls of the canisters, one of the methods more frequently used is the one called Summa® process. This method was originally developed in the seventies by R. A. Rasmussen and it consists in covering the inner surface of the canister with a layer of chrome-nickel oxide to reduce the number of sites in which adsorption processes can take place decreasing the reactivity of the canister walls. Besides the Summa® method, more recently it has been developed other procedure in which a thin layer of fused silica is bonded to the surface of the stainless steel.

The canister can be filled by the opening of a sampling port that allows the pass of air to the system (direct filling). The incoming flow of air has to be controlled using a flow-meter. In some occasions the canister is filled with a constant flow for a pre-established period of time in order to calculate daily averages (passive restricted sampling). The sampling can be also carried out in a pressurized way (active sampling) and in that case, although the sampled volume is higher than in the passive methodology, it does not mean that the samples are more concentrated. The inner pressure of the canister has to be among 15 and 20 psig (pounds per square inch gauge) and the pressure of 40 psig must not be exceeded in order to avoid risks during the transport. If the inner pressure of the canister is higher than the atmospheric pressure the canister can be evacuated in a "passive" way, whereas in the opposite case, the evacuation is developed by means of pumps. In some applications it is necessary to dilute the samples and, for this purpose, a flow of an inert gas has to be introduced in the canister to achieve the adequate inner pressure. Sometimes, this previous dilution of the sample implies posterior exhaustive concentration stages.

Anyway, it is very frequent that the samples taken using the canister methodology are concentrated previous to the injection in the chromatographic system by means of the trapping of the target analytes on an adsorbent bed or in a cryogenic trap. It is also possible to use a combination of both systems of preconcentration (see Table 1 for some examples).

Among the advantages of the VOC sampling using canisters the most important ones are the absence of formation of secondary products (after a correct cleaning stage previous to the sampling), the good suitability for non-polar analytes, the easiness of the handling and the fact of not having to supervise the overcoming of the breakthrough volumes of the target analytes. Besides these advantages, the use of canisters in the sampling stage is also compatible with analytes with a wide range of volatilities and high reactivities.

The main inconveniences of the sampling using canisters are the adsorption problems reported for polar analytes, the difficulties to achieve detection limits good enough for samples at low pressure, the requirement of carrying out the storage under controlled conditions of temperature and relative humidity, the high cost of the canister and of the additional instrumentation (as pumps, flow restrictors,...) especially when they are used sporadically, the problems with the storage and transport, their high weight and the difficulty of the automation of the sampling process. Besides this, the maintenance of the canister requires expensive and complex clean up procedures.

Additionally, it has been reported that some VOCs are not stable when the sampling is made using canisters. Thus, in the year 1995 Kelly and Holdren [107] had already reviewed the literature related to these problems. Later, other authors [108, 109] reported stability problems for the storage of biogenic VOCs in such receptacles. For example, Rhoderick [109] verified the isomerization of β-pinene into α-pinene, limonene, camphene and p-cymene, and the isomerization of α-pinene into camphene. Pallau and cols. [110] also detected losses of aromatic compounds as ethylbenzene and xylenes when the sampling was developed using a canister under conditions of low humidity. However, despite all these troubleshootings, the canister methodology has been widely used for the VOC sampling in air.

2.2. Cryogenic Trapping

Air sampling can be also made by cryogenic trapping. In this methodology the air sampled is forced to pass through a tube that can be empty or filled with an adsorbent material and this tube is kept at low temperature by means of its immersion in a cryogenic liquid as liquid nitrogen, CO_2 or argon. The low temperatures used in the cryogenic trapping can be also reached using electric devices but in these cases the temperatures achieved are not as low as the ones reached with the utilization of cryogenic liquids. At the same time, the use of adsorbent solids filling the cryogenic traps allows that the retention of the VOCs is produced at temperatures higher than the ones used in absence of adsorbent solids [111].

The adsorbents more frequently used in combination with the cryogenic trapping are glass beads, carbon derived adsorbents, synthetic polymers and molecular sieves.

Some examples of cryogenic trapping used for the VOC sampling in air are found in the literature [112-114].

Apart from the use for the direct sampling of VOCs, the combination of glass beads with nitrogen as cryogenic liquid is very utilized as a preconcentration system in the sampling with canisters, during the preconcentration stage previous to the chromatographic separation (see Table 1 for some examples).

The most important advantages of the use of cryogenic trapping for the VOC sampling in air are the simplicity of the devices and the achievement of very fast and complete desorptions. On the other hand, the principal drawback of these systems is that during the enrichment stage the caption of water and CO_2 is also favoured, and this may cause problems in the posterior chromatographic separation. In order to avoid such problems it is frequent to use molecular sieves or ascarite filters to trap these interfering substances. These filters are formed of an asbestos support covered with sodium hydroxide, although it has been reported that the use of such filters may also produce the retention of VOCs, limiting their application [111]. Other inconvenience derived from the use of cryogenic traps for the VOC sampling is that the utilization of cryogenic liquids to favour the condensation requires containers that occupy enormous volumes [115].

2.3. Adsorbent Solids

An adsorbent solid is a material that can retain molecules of other substances on its surface or in its interior by means of different physical and chemical interactions [37].

There is a wide variety of sampling devices for VOCs based on the use of adsorbent solids, especially for the analysis of VOCs with six or more carbon atoms or hydrocarbons with a smaller number of carbons atoms but with the presence of heteroatoms in their structures [111]. In fact, the EPA in the Method TO-17 [116] proposes the determination of VOCs in ambient air by active sampling using adsorbent solids.

For the sampling of lighter compounds it is preferable to use methods for the isolation of a whole volume of air or a cryogenic trapping method in order to assure the complete caption/retention of the analytes.

The sampling using adsorbents can be active or passive (by diffusion). In the active sampling the air is forced to pass through the adsorbent bed using a pump, whereas in the passive sampling the adsorbent is exposed to the air but without pumping mechanically the air through it.

The passive sampling is based on the formation of a concentration gradient through a diffusion barrier to produce a mass transfer of the gaseous molecules from the atmosphere to the adsorbent bed [39].

In the passive samplers (called diffusion samplers or dosimeters) the sampling rate is conditioned by the gaseous diffusion through an ideally static layer of air or a porous material and/or permeation through a membrane [117, 118]. Apart from this, the physical characteristics (length and diameter) and the chemical properties of the adsorbent also affect the diffusion rate. On the other hand, in the active sampling the collection rate depends on a flow rate previously fixed in the sampling pump.

Among the advantages of the passive sampling in comparison with the active one those more important are the easiness of the handling, the low cost, the lightness of the devices and their portability. Additionally, in the passive methodology it is not necessary to use a pump and this eliminates a potential source of contamination. The nature of the passive sampling also avoids the requirements of a power supply and batteries to keep the systems in good working order [119, 120].

The active sampling devices have the advantage of allowing the sampling of a higher air volume in a lower period of time than in the case of the passive methodologies, and this implies that the preconcentration of the sample is faster. Besides this, there are commercially available pumps with low weights and with long life batteries and this increases their portability.

As the passive samplers have a very easy operation mode and they are very light devices, these systems are frequently used for the personal sampling in occupational environments and in interior atmospheres in order to evaluate the levels of exposition to an specific pollutant that are suffered by some sectors of a working population. In fact, since these devices were introduced in the markets in the seventies they have found their major applications in this field [119, 121]. However, the use of passive samplers has also found application in the sampling of exterior atmospheres [122-124] although for this kind of applications the use of active sampling devices is more extended.

With reference to the geometry of the adsorbent traps used in the active sampling, they usually have a cylindrical shape. However, there are two kinds of devices to take samples of air in a passive way: on the one hand there are adsorbent traps with a cylindrical shape (tubes) and on the other traps with a plane shape (insignia samplers). In the samplers with a cylindrical shape the diffusion of the analytes is carried out in an axial path (through the length of the tube) while the transverse diffusion is smaller. In the samplers with a plane shape the longitudinal diffusion is very restricted and the transverse diffusion becomes more important. As a consequence of these differences in the directions of diffusion, the systems with a plane shape show higher caption rates than the cylindrical systems, although their plane geometry frequently makes them incompatible with the thermal desorption instruments used for the extraction of the analytes retained on the adsorbent beds [125].

Among the different models commercially available for the passive sampling of VOCs one of the most widely used is the one patented by the Salvatore Maugeri foundation under the name of Radiello®. This device consists of a tube filled with an adsorbent material suitable for the sampling of the target analytes and it is locked up in a porous polyethylene cartridge. In some of the applications that appear in the literature this tube is fixed on a triangular holder made of polycarbonate [126]. The most important innovation of this system in comparison with other models of passive samplers is that the diffusion path in the Radiello® sampler is parallel to the radius of the adsorbent tube, while in the conventional cylindrical passive samplers this diffusive path is parallel to the longitudinal axe. Other advantage of the Radiello® sampler in comparison with the remaining passive devices is the possibility of sampling analytes with high concentrations, in the range from 10^3 μg m^{-3} to 10^{10} μg m^{-3}, with a high caption rate without showing dependence on the air relative humidity and for wind speeds between 0.01 and 10 m s^{-1}. All these advantages make this system suitable for the VOC sampling in outside atmospheres [122].

Despite the fact that the Radiello® sampler is one of the systems with a more widespread use for the passive sampling of VOCs in air, there are other cheaper options that have demonstrated their versatility for the analysis of VOCs in different interior atmospheres as for example the VERAM (Versatile, Easy and Rapid Atmospheric Monitor) systems used by

Esteve-Turrillas and Ly-Verdu [127-129]. These samplers combine the use of adsorbent materials and semipermeable membranes in the sampling stage.

In order to evaluate if an adsorbent is suitable for the sampling of a specific category of organic analytes it is very frequent to use the calculation of the specific retention volumes and the calculation of the BreakThrough Volumes (BTV) for the target analytes. The specific retention volume is defined as the volume of a carrier gas that causes the migration of an analyte through an adsorbent bed with a mass of one gram at a given temperature. Whereas the BTV stands for the maximum volume of gas which can be sampled without dragging the retained compounds on a specific adsorbent mass, at a specific temperature [38]. The BTV depends on a great number of factors as the temperature, the concentration of the compound used in the experiments, the chemical composition of the gas that is under study, the relative humidity, the sampling flow of the gaseous current, the dimensions of the adsorbent trap, the mass and the granulometry of the adsorbent, the size of its porous, its specific surface, the number of cycles that the adsorbent has been reused and its thermal pretreatment. However, the parameters that affect the BTV to a higher extent are the temperature of adsorption and the concentration of the adsorbed compounds [130].

The air is a very complex matrix formed by hundreds of substances with different chemical and physical characteristics and also with different concentrations. Apart from this, the physical conditions of the air (temperature, humidity,) change constantly and they vary from one place to another. For these reasons it is obvious that there is no a single adsorbent with the appropriate characteristics for the sampling of all the VOC families and natures and, at the same time, valid for all the analytical applications [131]. In fact, there is a wide variety of adsorbents commercially available and the election of one of them depends on the kind and concentrations of the analytes of interest, on the nature of the matrix, and on the available techniques for the sampling and subsequent analyses.

In general terms, the most important characteristics that a good adsorbent solid has to compile are [38, 111, 131]:

- Complete enrichment of the analytes of interest. Frequently the value of the specific area (expressed in terms of surface units per mass units) and the characteristics of the porous surface can be useful parameters to make a first approach of the sorbent retention strength.
- The breakthrough volume has to be high enough to allow the sampling of large volumes of air without losses of the analytes of interest. The value of the BTV is also very useful in order to characterize the adsorbent strength from a quantitative point of view.
- Quantitative desorption of the analytes retained on its surface. The desorption stage has to be fast, simple and complete.
- Homogeneous and inert surface in order to avoid the formation of artefacts derived from the adsorbent, irreversible adsorptions or catalytic effects during the sampling time or the storage previous to the analysis.
- Low background signal and absence of interferences for the target analytes.
- Low water affinity in order to avoid the displacement of the retained compounds, the hydrolysis reactions and to minimize the possible alterations of the chromatographic process. The presence of water in the chromatographic system may produce the

deterioration of the stationary phase in the column and the modification of the retention times.

- Low ability to retain other inorganic air constituents as nitrogen oxides, SO_2, CO_2 or ozone.
- High chemical inertia towards reactive species as ozone.
- Total chemical inertia for the retained compounds, to avoid the decomposition of the constituents of the sample during the preconcentration and desorption stages.
- Good chemical and mechanical stability.
- High thermal stability.
- Possibility of reuse several times.

The adsorbent cartridges used for the VOC sampling can be filled with one single adsorbent (simple traps) or with a combination of several adsorbent solids (multisorbent or multicomponent traps).

Due to the difficulties to find a single adsorbent that fulfils the suitability requirements previously mentioned, the use of multicomponent traps is very usual for the VOC sampling in air since the combined use of different adsorbents allows the retention of compounds with different physicochemical characteristics.

In the multicomponent traps each adsorbent is separated from the rest by plugs of glass wool or metallic grates with an adequate mesh to avoid the mechanical dragging of the small particles of the adsorbent material. In the preparation of such kind of traps the adsorbents have to be placed following an increasingly or decreasingly order according to their specific areas and, therefore, a specific order in their adsorption strengths. Thus, during the sampling stage the adsorbent with the lowest specific area (the weakest adsorbent) has to be the first one contacting with the sampled air. In this way the following adsorbents, with a higher sorbent strength, are protected from irreversible adsorptions avoiding, at the same time, their saturation. The extraction stage in these multicomponent traps has to be developed in the inverse order to avoid that those analytes with the lowest volatilities, retained on the adsorbent with the smallest specific area, go through the remaining adsorbent beds. Thus, those VOCs with the lowest volatilities will be never in contact with the strongest adsorbents.

Some of the advantages derived from the use of adsorbent cartridges in the sampling stage are the easiness of the manipulation, the small size and the relative low cost [132]. Moreover, as it was mentioned before, this sampling system offers the option of combining different kinds of adsorbents in a single cartridge to retain compounds with different natures belonging to the same sample. The sampling by means of adsorbent solids can be automated and programmed in a sequential mode of operation allowing the unattended monitoring of VOCs during long periods of time.

Other advantage that justifies why the use of adsorbent cartridges is so extended for the VOC sampling in air is because they can be reutilized several times subjecting them to a cleaning stage before each individual use.

Additionally, the adsorbents are quite selective for one or various categories of compounds, and this allows that some undesirable interferences can be eliminated even in the sampling stage; for example Roukos and cols. [133] used an adsorbent trap formed by two hydrophobic graphitized carbons (Carbopack B®:Carbopack X®) in order to minimize the effect of water vapour in the sampling of oxygenated volatile organic compounds in air.

Borusiewicz and Zieba-Palus [134] verified that in the VOC sampling using Tenax TA® the nonpolar and the heaviest compounds displace volatile and polar compounds while Carbotrap® was demonstrated to be more effective for the sampling of volatile and polar compounds. Gallego and cols. [135] proved that multisorbent traps formed by Carbotrap®:Carbopack X®: Carboxen 569® showed a better performance than Tenax TA® traps for very volatile compounds, especially for compounds with boiling points lower than 100°C and pressures of vapour higher than 3kPa (20°C). Dettmer and cols. [136] demonstrated that some very light compounds as isoprene and 1,3-butadiene are not completely retained on some carbon molecular sieves (for example Carboxen 569®, Carboxen 1003® and, especially, Carbosieve SIII®) whereas other molecular sieve, Carbotrap X®, is not suitable for the retention of propene and propyne.

Additionally, the use of adsorbent solids has other advantage since during the sampling stage it is produced the enrichment in the target analytes on the adsorbent surface and this allows eliminating a subsequent stage of preconcentration of the samples.

The main disadvantage of the VOC sampling with adsorbent solids is that some adsorbents can react with oxidant substances of the air or they can simply alter their compositions releasing different artefacts. This deterioration produces the alteration of the capacity of these substances to retain the target compounds and it also causes the appearance of undesirable interferences in the chromatographic separation. Apart from this, it must be taken into consideration that the exposure to the atmospheric air oxidants not only affects the adsorbent solids but also the organic analytes retained on the adsorbent bed. For example, the degradation of the polymer Tenax® produces the liberation of VOCs as benzene, toluene, benzaldehyde, phenol, acetophenone, linear-chain aldehydes (hexanal, heptanal, octanal or nonanal), 2,4-diphenyl-p-benzoquinone, 2,6-diphenyl-p-hydroquinone and 2,6-diphenyl-p-quinol [137]. Besides this, it has been reported [138] that some compounds like isoprene can suffer isomerizations when they are retained on Tenax TA® traps. One of the most important problems derived from the reaction and ulterior decomposition of the adsorbent materials with oxidant substances of the air is that the formation of artefacts is not proportional to the sampled volume of air, and, for this reason, it is impossible to know how much of the compound comes from the sampled air and which percentage is produced by the reaction of the adsorbent material [139]. To avoid the degradation both of the adsorbent and of the retained analytes, it is possible to use a scrubber or a system to retain the oxidant substances in the incoming flow of air before its entrance in the adsorbent cartridge [140-142]. Some of the different systems to eliminate the oxidants in the flow of air are tubes filled with KI crystals, metallic copper in tubes covered with KI or combined with MnO_2 and $FeSO_4 \cdot 7H_2O$ [140].

Other disadvantage derived from the use of adsorbent solids for the sampling is that not all the compounds can be retained on these supports, since some compounds as mercaptans and reduced compounds containing sulphur as CS_2 or H_2S are not stable on the most of the adsorbent solids [143]. Besides this, some compounds extremely volatile as the ethane, ethylene or propylene cannot be retained on adsorbent solids without using a cryogenic trap in the sampling stage. Consequently the use of adsorbents is, in a sense, a discriminatory sampling method and it avoids obtaining the qualitative composition of all the analytes in the sampled air volume.

Apart from this, it is also important to take into account that not all the adsorbents show hydrophobic properties and, for this reason, if a purge stage is not included in the analytical procedure, problems in the posterior chromatographic separation and detection can emerge. The moisture can be also eliminated by means of a device that captures it before the air reaches the adsorbent tube. One option is to use Nafion® membranes [34, 144, 145] which allow the reduction of the moisture up to 20 times when they are recently conditioned. However, among the inconveniences of using Nafion® membranes it is remarkable that some monoterpenes (for example α-pinene or β-pinene) can suffer from structural changes when they are retained on this material and, at the same time, some hydrocarbons can be selectively retained on the surface of this polymer. Other option to retain the moisture is to use a moisture precipitator that produces the water condensation by means of compression or cooling procedures. The main drawback of such systems is that they can also produce losses of polar compounds retained on the aqueous phase. The moisture can be also eliminated using desiccant substances as $CaSO_4$, $CaCl_2$, $Mg(ClO_4)_2$ and K_2CO_3 [146, 147] which show more versatility than the previous techniques. All these methods to retain the moisture can be also combined with the sampling of a whole air volume.

Other drawback of the sampling using adsorbent materials is that when the extraction of the analytes retained on the surface of the material is carried out by thermal desorption each sampling is unique, and for this reason is not possible to analyze several aliquots of the same sample. In order to solve this problem some commercial trademarks have developed thermal desorption instruments that allow the recirculation of the gas outgoing through the splits valves towards the adsorbent trap in order to analyze a second aliquot of the same sample.

The use of adsorbent solids in the sampling stage produces also elevated background signals on the analytical method although this contribution to the total backgrounds can be minimized incorporating exhaustive cleaning stages before the sampling. It is indispensable to evaluate the background signal derived from the use of adsorbent materials, especially when the traps are reused several times and to change the adsorbent solid before the background signal is very high. As it was mentioned before, to use an adsorbent cartridge for the first time it is necessary to subject the adsorbent to a conditioning stage and this stage can last for several hours. Before the subsequent uses, the adsorbent beds are subjected to a cleaning stage that usually lasts for a few minutes. The initial conditioning of the adsorbent beds usually consists in subjecting the adsorbent to temperatures close to its maximum stability temperature under a flow of an inert gas in order to eliminate impurities from the surface of the adsorbent solid. In those cases in which the development of the analytical method is based on the use of solvents for the extraction, the initial conditioning of the adsorbent solids can be also carried out washing the material with appropriate solvents. Sometimes the use of solvents for the cleaning stage is also combined with an ulterior thermal desorption stage. Anyway, both the initial conditioning and the cleaning stage previous to each sampling produce an increase of the analysis time.

Besides all the inconveniences cited above, the use of adsorbent solids for VOC sampling is a technique that does not allow the sample analyses in real time, although nowadays there are commercially available some automatic systems in which one sample is being taken on an adsorbent cartridge while other tube is being simultaneously analyzed [144, 145, 148].

The limitation derived from the delay in the analysis is common for the VOC sampling using adsorbent solids, cryogenic trapping and whole air sampling. However, as it was explained before, there are other analytical techniques in which the flow of sampled air is directly driven into the instrumental equipment to produce an analytical response in real time. This is the case, for example, of the analytical systems based on the use of Fourier Transform InfraRed (FTIR)spectroscopy which allows the VOC continuous monitoring in real time [149].

The solid materials used for the sampling of VOCs in air can be classified into three major categories: inorganic materials, carbon derived adsorbents and organic polymers.

2.3.1. Inorganic Materials

In this group of substances there are different materials as silica gel, zeolites, Florisil® or alumina. In general terms, the inorganic adsorbents are not used for the sampling of low concentrations of VOCs in environmental monitoring programmes because of their high hydrophilic properties. In fact, their hydrophilia restricts the application of these sorbents in this field and, in fact, this kind of adsorbents has traditionally found more applications in the elimination of harmful substances in the industrial hygiene area. For example, the silica gel is used for the sampling of alcohols, amines and phenols on a large scale. One of the inconveniences derived from the use of this adsorbent is that it has a big specific surface and a marked polarity and, for this reason, this adsorbent shows a marked preference to retain water molecules. Consequently, this can produce the displacement of other retained molecules which can be interpreted as an apparent overcoming of the breakthrough volume [150]. Zeolites are widely used in an industrial scale to eliminate some solvents from the air as light alkanes and ketones. These adsorbents show the advantage of being stable under high temperatures, keeping a high adsorbent strength under conditions of moderate humidity. Besides this, these adsorbents are not inflammable [151].

2.3.2. Carbon Derived Adsorbents

The carbon derived adsorbents can be classified into activated carbons, carbon molecular sieves, graphitized carbon blacks and porous carbons [136].

2.3.2.1. Activated Carbon

Activated carbon is a microporous substance that can be obtained from a great variety of carbon-based materials as biomass residues (wood, coconut shell or fruit pits), fossilized matter (peat, lignite and any other kinds of charcoal) or organic polymers [152].

The manufacturing of this kind of substances consists of a carbonization stage and an activation phase to achieve the partial gasification of the coal. The gasification process is responsible for the porous structure of the coal and therefore this stage conditions the properties of the carbons and the high specific surface of such substances (in the range of 300 and 2000 m^2 g^{-1}). Apart from these characteristics, the activated carbons also show an elevated thermal stability (>700°C).

The adsorption related properties of this kind of substances are a consequence of their high specific areas, of the presence of microporous in their structure and of the wide variety of functional groups that they have. In the carbonaceous matrix there may be different functional groups with heteroatoms as oxygen, nitrogen or sulphur. These functional groups condition the characteristics of the adsorption and the reactivity of the adsorbent. Besides this, there are several techniques as heating treatments, oxidation and refining processes and also addition of inorganic substances that modify the reactivity of the surface and the structural and chemical properties of the carbon.

The adsorption of the compounds on the surface of the activated carbon is mainly controlled by two different kinds of interactions: physical interactions, including the exclusion by size and the effects of microporosity, and chemical interactions which are conditioned by the chemical nature of the adsorbent and also by the nature of the adsorbate [153].

One of the major applications of the activated carbon is the elimination of VOCs in effluents at an industrial scale. Despite the fact that this method is very widespread it has a main drawback due to the exothermic nature of the adsorption process. This phenomenon is more important as the VOC concentrations in the gaseous stream are elevated. Taking into account that the carbon is a flammable material, in such applications there is a real risk of fire [154].

The use of activated carbon for the sampling of VOCs has been focused on the field of the industrial hygiene. In fact, there is a great number of official analytical methods for the analysis of organic substances in interior air based on the use of activated carbon for the sampling. Activated carbon shows a good ability for the adsorption of hydrocarbons C_2-C_6, vinyl chloride, esters, ethers, alcohols, glycol ethers and halogenated hydrocarbons. However, those compounds with a lower volatility and a major reactivity as amines, phenols, nitrogen compounds or aldehydes show poorer recoveries, since these categories of compounds frequently suffer irreversible adsorptions. The microporous structure of the activated carbon makes it difficult to recover compounds with a low polarity as benzene, toluene or xylenes. Besides this, some inorganic compounds as NO_2, Cl_2, H_2S, SO_2 and ozone react with the activated carbon to produce interfering compounds that alter the adsorption properties of this material.

Most of the problems associated with the use of activated carbon in the analytical applications derive from the variations in the physical and chemical properties of this adsorbent. Besides this, some compounds can be so strongly retained that the ulterior extraction cannot be complete. Other of the inconveniences associated with the use of activated carbon is the enormous difficulties to obtain low background levels in comparison with the ones produced by other kind of adsorbents. Additionally activated carbon also shows a marked hydrophilia [37].

It is frequent that the utilization of activated carbon for the VOC sampling is combined with the posterior solvent extraction producing a dilution of the samples and making the detection limits worse. The combination of activated carbon with the solvent extraction is very frequent in the analysis of interior air [155].

2.3.2.2. Molecular Sieves

Molecular sieves are produced by the controlled pyrolysis of an adequate polymeric material as polyvinyl chloride and polyvinylidene chloride or petroleum resins (bitumen) at temperatures usually higher than 400°C. This kind of adsorbents is characterized by their porous structure with practically uniform porous, specific areas in the range from 400 to 1200 m^2 g^{-1} and by their thermal stability even at temperatures higher than 400°C. The structure, the purity and the technology used to carbonize the base material condition the mesh size, the number, size and distribution of the porous in the final product. The presence of impurities in the polymeric matrix produces a drastically decreasing in the homogeneity of the porous structure [37].

These adsorbents are suitable for the recovery of very volatile compounds, with boiling points in the range from -15°C to 120°C, and relatively nonpolar compounds as benzene, toluene, methyl chloride or vinyl chloride [155]. In this category of carbons there are some commercial marks as Carbosieve SIII®, Carbosieve G®, Carboxen 563®, Carboxen 564®, Carboxen 569®, Carboxen 1000®, Carboxen 1004®, Spherocarb®, Carbosphere® and Ambersorb®.

2.3.2.3. Graphitized Carbon Blacks

Graphitized carbon blacks are substances with a homogeneous surface and without microporous in their structures. The adsorption takes place in the external surface of the carbon and it is conditioned by the size and shape of the adsorbed molecules. In this kind of materials the processes of sorption based on hydrogen bonds are limited and they are not sorbents suitable for very small and polar molecules.

In fact, one of the most remarkable properties that makes the use of this kind of substances very attractive is that they show low affinity towards the water, although the modification of the chemical structure of the carbon by means of oxidation processes can alter this property.

The synthesis of the graphitized carbon blacks begins with a heating at 3000°C of black carbons produced by the incomplete combustion of products derived from the carbon. This treatment eliminates the volatile compounds and the tarry residues and, at the same time, it induces the growing of small graphite crystallites that are particles in which the graphite crystals are arranged in polyhedrons [152]. At the same time, some functional groups of the carbon surface are eliminated. The greater graphitization grade the lower number of residual porous in the surface of the carbon and, therefore, the small specific area (between 5 $m^2 g^{-1}$ and 250 $m^2 g^{-1}$). The graphitized carbon blacks are thermally stable up to temperatures higher than 400°C.

Some examples of graphitized carbon blacks are Carbotrap®, Carbotrap B®, Carbotrap C®, Carbopack®, Carbopack B®, Carbopack C® and Carbopack F®. The difference between the Carbopack® and Carbotrap® series is the mesh size of the adsorbents. Thus, Carbotrap® adsorbents are 20/40 mesh while Carbopack® adsorbents are 40/60 mesh or smaller.

2.3.2.4. Porous Carbons

Porous carbons are materials with a homogeneous surface, with hydrophobic properties, high mechanical resistances and also with a high stability in a wide range of pH.

This kind of carbons is synthesized impregnating a silica gel support that acts as a template with an organic precursor (saccharose, resins formed by phenol and formaldehyde or resins of phenol and hexamine).

After the polymerization in the porous of the material used as template, the polymer is subjected to 1000°C under an inert atmosphere and it becomes a glassy carbon. After the pyrolysis, the silica gel support is eliminated by means of an alkaline hydrolysis to get a material whose particle size and shape, number and size of porous and specific area are conditioned by the initial characteristics of the template used [156]. The last stage consists in the heating of the material obtained previously, under an inert atmosphere and temperatures in the range from 2000 to 2800°C in order to anneal the surface, eliminate the porous and, depending on the temperature, to get some grade of graphitization [152].

2.3.3. Synthetic Organic Polymers

Synthetic organic polymers are a wide group of substances with different specific areas and polarities. Some of these polymers have their origins in the stationary phases of the packed chromatographic columns. As a consequence of the control of the conditions during the manufacturing process, the porous polymers are usually very pure substances. One of the main drawbacks showed by this category of sorbents is that they have lower thermal stabilities than the carbon derived adsorbents and, for this reason, some of them cannot be extracted by thermal desorption techniques.

Some of the polymers most widely used for the sampling of VOCs in air are Tenax TA®, Tenax GC®, Tenax GR®, Chromosorb® series, Porapak® series, HayeSep® series and amberlite resins as XAD® resins. In the successive sections some of these families of substances will be presented.

2.3.3.1. Tenax®

Tenax® (polyoxide-2,6-diphenyl-p-phenylene) is one of the polymeric materials most widely used for the sampling of VOCs in air. The ability of this polymer to concentrate VOCs with different natures and with a wide range of boiling points makes it indispensable for the analysis of samples with a complex and unknown nature [157]. The firstly developed polymer of this family was the Tenax GC® characterised by a very small specific area (19-30 $m^2\ g^{-1}$), a high thermal stability (450°C) and a low affinity towards the water. It is a polymer recommended for the sampling of compounds with an intermediate volatility (C_6-C_{14}) showing the advantage of its marked hydrophilicity. Tenax TA® has a specific surface slightly higher (35 $m^2\ g^{-1}$) than Tenax GC® although its thermal stability is lower (300°C). The background signal produced by Tenax TA® is lower than the one produced by Tenax GC® since Tenax TA® shows a lower bleeding. There is a third variety of Tenax®: Tenax GR®. This polymer has a base of Tenax TA® and a 23% of graphitized carbon. The presence of the graphitized carbon in its structure allows the adsorption of lighter molecules but it increases the quantity of water that it may retain.

2.3.3.2. Chromosorb®

The adsorbents of the Chromosorb® series are obtained by the polymerization of styrene and divinylbenzene with the possible presence of other molecules as acrylonitryle. The adsorbents belonging to this series are suitable for the retention of light and small molecules.

2.3.3.3. Porapak® and Hayesep® Series

The adsorbents from the Porapak® series are polymers of pyridine, pyrrolidine and dimethylbenzene, while the members of the series HayeSep® are polymers and copolymers of divinylbenzene and other compounds.

Both series are valid for the sampling of mixtures with oxygenated and halogenated substances (water, alcohols, glycols, ketones, aldehydes, fatty acids and esters).

2.3.3.4. Amberlite Resins: XAD®

This family of adsorbents is formed by non-ionic reticulate resins with thermal stabilities lower than 200°C. For that reason they are usually extracted using solvents.

In the Table 2 the characteristics of some commercially available adsorbents are summarized.

Table 2. Characteristics of some common adsorbents used for the VOC sampling

Nature		Adsorbent	Range of analytes[(1)]	Maximun temperature (°C)	Specific surface (m^2g^{-1})
Inorganic adsorbents		Glass fiber	from n-C_{20} to n-C_{36}	400	
Carbon derived sorbents					
	Activated carbon	Carbon	bp. from -80°C to +50°C	400	>1000
	Molecular sieves	Spheropcarb®	bp. from -60°C to +80°C	400	1200
		Carbosieve SIII® Carboxen® series	bp. from -30°C to +150°C	400	800
	Graphitized carbon blacks	Carbotrap C® Carbopack C® Carbopack Y®	from n-C_8 to n-C_{30}	400	12
		Carbotrap® Carbopack B®	(n-C_4) n-C_5 to n-C_{14}	400	100
Synthetic organic polymers		Tenax TA®	n-C_7 to n-C_{26}	350	35
		Tenax GR®	bp. from 100 to 400°C from n-C_7 to n-C_{26}	350	35-100
		Chromosorb 102®	bp. from 50 to 200°C	250	350
		Chromosorb 106®	bp. from 50 to 200°C	250	750
		Porapack Q®	bp. from 50 to 200°C from n-C_5 to n-C_{12}	250	550
		Porapack N®	bp. from 50 to 150°C from n-C_5 to n-C_8	180	300

(1) bp.: boiling point.

Figure 1 shows some images of several adsorbents obtained by scanning electron microscopy. In these images the morphological differences among them can be appreciated. The particles of Carbotrap C® (Figure 1.A) practically do not show porous in their surface and, for that reason, the specific surface of this adsorbent is quite small (~12 g m^{-2}). Chormosorb 106® (Figure 1.B) also shows a structure with a small number of porous, however this adsorbent has a higher specific surface (~750 g m^{-2}) than the previous adsorbent. This may be a consequence of the differences in the shape of the particles that form these adsorbents. Thus, in the case of Carbotrap C® the particles are morphologically very similar whereas Chromosorb 106® is formed by particles with different shapes (spheres, semi-spheres and porous particles) that confer more specific area to this last adsorbent. Regarding Tenax GR® (Figure 1.C) and Tenax TA® (Figure 1.D) both adsorbents show a structure with a major number of porous than the Carbotrap® and, therefore, a higher specific area. Besides this, the graphitized carbon in the structure of Tenax GR® gives more porosity to its particles justifying the higher specific surfaces (35-100 g m^{-2}) for this adsorbent in comparison with the Tenax TA® (35 g m^{-2}).

In the Table 3 some applications related to the use of adsorbent solids in the sampling stage are shown.

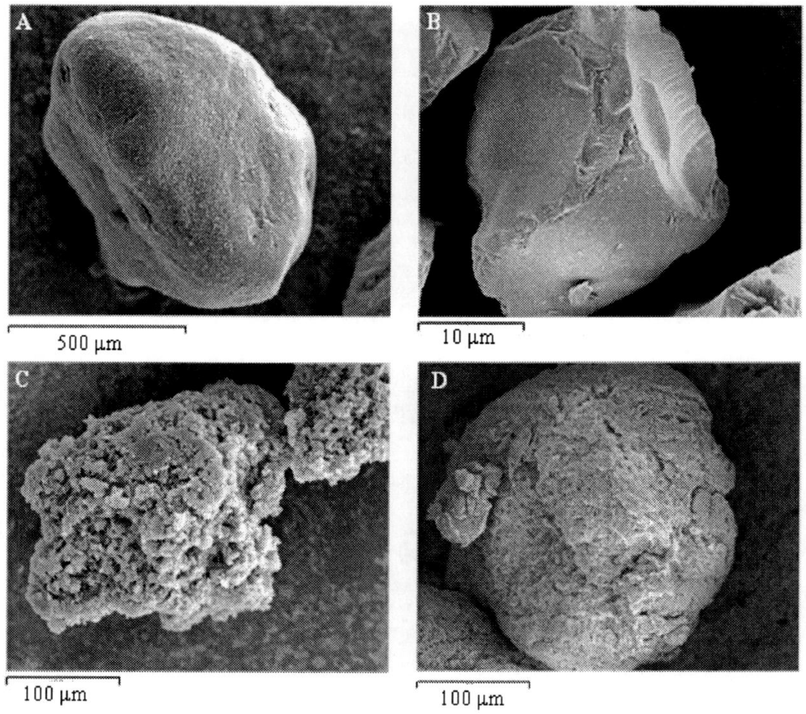

Figure 1. Images obtained by scanning electron microscopy from particles of different adsorbents: Carbotrap C® (A), Chromosorb 106® (B), Tenax GR® (C) and Tenax TA® (D).

Table 3. Examples of adsorbent enrichment for VOC sampling

Sampling(1)	**Extraction**	**Ulterior analysis**(2)	**Application**(3)	**Citation (year)**
Tenax GC®	Thermal desorption	GC/FID	Aromatic and halogenated VOCs in ambient air	[158] (1984)
Tenax GC®	Thermal desorption	GC/FID	BTEX and aliphatic VOCs in ambient air sampled in London	[159] (1994)
Activated carbon	CH_2Cl_2	GC/FID	BTEX and light PAHs in ambient air in Greece	[160] (1994)
Activated carbon	Thermal desorption	Cryoconcentration/ GC/FID	NMHC in ambient air in Sweden	[161] (1994)
Charcoal (OVM 3000) (Passive sampling)	CS_2	Dual GC/ECD/FID	Monitoring of 28 VOCs in outdoor and indoor air in an industrialized city and in an urban area in Germany	[162] (1996)
HayeSep D®: Anasorb 747® (cylindrical denuder sampler)	----	GC/PID	Determination of the adsorption isotherms of n-hexane in a denuder sampler using different models	[163] (1996)
Super Q®	Solvents	GC/FID GC/MSD	BVOCs released by plants as a consequence of the attack produced by insects	[164] (1996)

Sampling(1)	Extraction	Ulterior analysis(2)	Application(3)	Citation (year)
Tenax TA® and carbonaceous sorbents	Thermal desorption	GC/FID GC/MSD	BVOCs in ambient air (interlaboratory exercise)	[165] (1997)
Carbosieve SIII® and Carbotrap®	Thermal desorption	GC/FID	Characterization of the terpenoid emissions by plants in the Mediterranean area	[166] (1997)
Tenax GR® Carbosieve SIII®	Thermal desorption	Cryoconcentration/ GC/FID	BVOCs emitted by different *Eucalyptus* species	[167] (2000)
Carbotrap B®: Carboxen 1000®: Carboxen®	Thermal desorption	GC/FID/ELCD	Measurements of aromatic and halogenated VOCs in an urban atmosphere in Yokohama (Japan)	[168] (2000)
Tenax TA®: Carbotrap®	Thermal desorption	GC/FID GC/MSD	BVOCs in a coastal area of Ireland	[169] (2001)
Chromosorb®	Thermal desorption	GC/MSD	Grass-atmosphere gaseous interchange (benzene, toluene and halogenated VOCs)	[170] (2002)
Carbograph 4® (Passive sampling)	Thermal desorption	GC/FID	Monitoring of VOCs in ambient air in Madrid (Spain)	[122] (2002)
Tenax TA® filling a liner of a gas chromatograph	Thermal desorption	GC/MSD	Determination of benzene and other VOCs emitted in the combustion of charcoal	[183] (2003)
Tenax GR®	Thermal desorption	GC/FID (portable)	*In situ* determination of VOCs in an office building in Ohio, EE.UU.	[184] (2003)
Tenax TA®: Carboxen GR®	Thermal desorption	GC/MSD	Determination and quantification of VOCs that provide odour around a dairy in California	[185] (2003)
Tenax TA®	Thermal desorption	GC/FID/MSD	Study of the effect of the exposure to VOCs in several urban populations from different European cities	[186] (2003)
Activated carbon	CS_2	GC/MSD		
Activated carbon (Passive sampling)	CS_2	GC/FID	Determination of benzene in different working ambients in Copenhagen	[187] (2003)
Glass beads at -100ºC	Thermal desorption	Bidimensional GC/MSD	On-line analysis of cigarette smoke	[188] (2003)
Carbopack C®: Carbopack B®: Carbosieve SIII®	Thermal desorption	GC/MSD	Measurement of VOCs in the interior of semiconductor foundries	[189] (2003)
Tenax TA®: Carbosieve SIII®	Thermal desorption	GC/MSD	Study of the influence of the cigarette smoke on the VOC concentrations measured in air. Study of the relations of such concentrations with the tobacco smoke tracers	[190] (2003)
Tenax TA®: Carbograph®	Thermal desorption	GCxGC/FID	Optimization of a method of GCxGC bidimensional for the analysis of VOCs in the atmosphere	[191] (2003)
Tenax TA®: Carbopack B®:Carboxen 1000®	Thermal desorption	GCxGC/TOF-MSD	Optimization of a method of GCxGC bidimensional for the analysis of VOCs in the atmosphere. Identification of compounds	[191] (2003)
Tenax® Carbosieve SIII® Silica gel	Thermal desorption	GC/MSD	Characterization of VOCs in a landfill site in China	[192] (2003)
Carbopack B®: Carbosieve SIII®	Thermal desorption	GC/MSD	Study of the effect of the ventilation systems in buildings with conditioned air in Singapore	[193] (2003)
Carbopack B®: Carbosieve SIII®	Thermal desorption	Bidimensional GC/FID	Determination of VOCs in ambient air during a four-year monitoring campaign in France	[34] (2004)
Carbotrap C®: Carbotrap®: Carbosieve SIII®	Thermal desorption	GC/MSD	Characterization of VOCs in tobacco smoke	[194] (2004)

Table 3. (Continued)

Sampling(1)	Extraction	Ulterior analysis(2)	Application(3)	Citation (year)
Carbotrap®: Carbosieve SIII®	Not-specified	GC/MSD	Study of the implication of terpenes in the formation of secondary organic aerosols	[195] (2004)
Tenax GR®: Chromosorb 106®	Thermal desorption	GC/MSD	Measurement of the concentrations of C_1-C_9 hydrocarbons in urban air of the city of A Coruña (Spain)	[196] (2004)
Tenax GR®	Thermal desorption	GC/FID	Release of VOCs emitted by carpets	[197] (2004)
Tenax TA®	Thermal desorption	GC/MSD	Study of the levels of VOCs derived from the tobacco smoke	[198] (2004)
Carbopack B®: Carbotrap C®: Carboxen 1000®	Thermal desorption	GC/MSD	*In situ* analysis of VOCs. Comparison of methodologies	[113] (2004)
Al_2SO_3 at -160°C	Thermal desorption	GC/FID		
Tenax TA®	Thermal desorption	GC/MSD	Measurement of VOCs in interior, exterior and in environmental working air	[199] (2004)
Graphitized carbon black: carbon derived molecular sieve	Thermal desorption	GC/FID	Study of the relation of submicrometer particles and VOCs using receptor models	[200] (2004)
Not specified	Thermal desorption	GC/FID	Determination the levels of exposure to VOCs suffered by children attending class	[201] (2004)
Carbon derived (Passive sampling)	Acetone:CS_2 (2:1, v/v)	GC/MSD	Determination of the personal exposure levels in the urban ambient of Minneapolis	[64] (2004)
Activated carbon (Passive sampling)	Toluene	GC/ECD	Comparison of the halogenated VOC levels in Japan and Sweden both in interior air and in ambient air	[202] (2004)
Activated carbon (Passive sampling)	CS_2	GC/FID/ECD	Determination of VOCs in indoor air sampled in different rooms of houses located in several German cities	[203] (2004)
Carbon derived (Passive sampling)	Acetone:CS_2 (2:1, v/v)	GC/MSD	Determination of VOCs in indoor and exterior air and measurement of the personal exposure to VOCs in three communities in Philips (Minneapolis, EE.UU.)	[123] (2004)
Activated carbon	CS_2	GC/FID or GC/MSD	Study of the effect of the reduction of the benzene contents in gasoline on the values of BTEX in air in Toulouse (France)	[204] (2004)
Activated carbon derived of coconut shell	Thermal desorption	GC/MSD	Determination of VOCs in ambient air in Mumbai (India) to validate a receptor model	[205] (2004)
Carbopack C®: Carbopack B®: Carbosieve SIII®	Thermal desorption	GC/MSD	Measurements of VOCs in the entrance and exit of the air condition systems to model the emission trends	[206] (2004)
Tenax GC®	Thermal desorption	GC/FID	Determination of the pollutant levels emitted by vehicles in Taipei (China)	[207] (2004)
Carbotrap C®: Carboxen 564®; Carboxen 1016®; Carboxen 1016®: Carbotrap C®	Thermal desorption	GC/MSD	Determination of halogenated VOCs in gaseous samples taken during a research cruise from Germany to Antarctica and in Ireland. Comparison of different adsorbent traps	[208] (2004)
Activated carbon (Passive sampling)	CS_2	GC/FID	BTX analysis in several points of Tokyo (Japan)	[209] (2004)

Sampling(1)	Extraction	Ulterior analysis(2)	Application(3)	Citation (year)
Tenax TA®: Carbosieve SIII®	Thermal desorption	GC/MSD	Determination of the exposure levels to VOCs in different streets of Guangzhou (China)	[210] (2004)
Tenax®: Ambersorb®: Carbon: Carbotrap 300®	Thermal desorption	GC/MSD	Determination of benzene in several locations of New York (EE.UU.) between 1990 and 2003. Comparison of different methodologies for VOC sampling (canister versus adsorbents)	[68] (2005)
Tenax TA®: Carbopack®	Thermal desorption	GC/FID	Development of a model to describe the temporal fluctuations of monoterpenes produced by Scots pine using an enclosure system	[211] (2005)
Tenax TA®	Thermal desorption	GC/FID	Characterization of the monoterpene emissions produced by *Fagus sylvatica* in a French forest	[212] (2005)
Tenax GC®: Carbosieve®	Thermal desorption	GC/FID	Determination of the emission fluxes produced by tropical plant species in India using an enclosure system	[213] (2005)
Tenax GR®: Tenax TA®: Carbotrap C®	Thermal desorption	GC/FID and GC/MSD	Determination of the sesquiterpene emissions produced by *Pinus Taeda* L. and study of the relation with the aerosol formation	[214] (2006)
Tenax TA®: Cartbotrap®	Thermal desorption	GC/MSD	Daily and seasonal evolution of the mono and sesquiterpenes emitted by *Pinus sylvestris* L.	[215] (2006)
Carbon molecular sieves	Thermal desorption	GC/FID	Study of the influence of isoprene concentrations and other VOCs on the O_3 formation in Taiwan	[216] (2006)
Carbotrap C®: Carbotrap®	Thermal desorption	GC/MSD	Determination of the factors that have an influence on the linalool and ocimene emission by *Pinus Pinea* L.	[217] (2006)
Tenax TA®	Thermal desorption	GC/MSD	Determination of the VOC sources in an Alpine valley in Chamonix	[218] (2006)
Tenax TA®	Thermal desorption	GC/FID	Determination of the isoprene fluxes produced by several tropical tree species in Okinawa Island (Japan) using an enclosure system	[103] (2006)
Carbopack C®: Carbopack B®: Carbosieve SIII®	Thermal desorption	GC/MSD	VOC source identification in buildings with air conditioned systems in Singapore. Comparison with European buildings	[219] (2006)
Carbotrap C®: Carbotrap B®: Carbosieve SIII®	Thermal desorption	GC/MSD	Study of the interchange fluxes of CO_2 and monoterpenes between the soil and the atmosphere under drought conditions	[220] (2007)
Carbotrap®: Carbopack X®: Carboxen 569®	Thermal desorption	GC/MSD	Development and validation of a method for VOC analysis. Application to the analysis of urban and industrial air samples	[221] (2007)
Tenax TA®	Thermal desorption	GC/MSD	Estimation of the potential emission of BVOCs in the forest and in the paddy fielding in the Kinki region (Japan)	[222] (2008)
Carbopack B®: Carbopack X®. Carboxen 101®	Thermal desorption	GC/MSD	Study of the influence of the garages and basements on the VOC levels measured in the interior of houses	[223] (2008)
Carbon derived	Thermal desorption	Bidimensional GC/FID	Determination of the VOC levels in a natural park of the Basque Country (Spain)	[145] (2008)
Carbotrap®: Carboxen 1003®	Thermal desorption	GC/MSD	Development of an automated system for the analysis of VOCs in air using an on-line internal standard. The adsorbent trap is used to concentrate the samples directly in the analytical instrument	[224] (2008)

Table 3. (Continued)

Sampling[(1)]	Extraction	Ulterior analysis[(2)]	Application[(3)]	Citation (year)
Carbon derived molecular sieve	Thermal desorption	Bidimensional GC/FID	Determination of VOCs in urban and rural air in China	[225] (2008)
Carbon derived molecular sieve	Thermal desorption	Bidimensional GC/FID	Relation between VOCs and ozone in Pearl River Delta (China)	[226] (2008)
Activated carbon (Passive sampling)	CS_2	GC/MSD	Determination of BTEX in an area subjected to heavy road traffic	[227] (2009)
Carbon derived adsorbent (Passive sampling)	CS_2:acetone	GC/MSD	Determination of VOCs after the Katrina hurricane in New Orleans (EE.UU.)	[228] (2009)
Activated carbon (Passive sampling)	CS_2	GC/MSD	Comparison of the indoor air quality in Hangzhou (China) and in Shizuoka (Japan)	[229] (2009)
Tenax TA®	Thermal desorption	GC/MSD GC/FID		
Carbosieve SIII®: Carbotrap B®: Carbotrap C®; Tenax TA®	Thermal desorption	GC/MSD	Determination of the isoprene and monoterpenes emission fluxes produced by the tree species *Quercus robur* using an enclosure system	[138] (2009)
Tenax TA®	Thermal desorption	GC/MSD	Study of the evolution of the BTEX concentrations in an Atlantic suburban area of Spain during a year	[230] (2009)
Carbopack B®: Carbopack X®	Thermal desorption	GC/FID	Development and application of a method for the automatic monitoring of oxygenated VOCs and nitrile compounds in air	[133] (2009)
Carbotrap C®: Carbotrap®: Carbopack X®	Thermal desorption	GC/MSD	Characterization of the BVOC fraction released by different eucalyptus species in Australia	[231] (2009)
Tenax TA®: Carbopack B®	Thermal desorption	GC/MSD	Study of the BVOC emissions, excluding isoprene, in boreal peatlands	[232] (2010)
Tenax TA®; Carbotrap®: Carbopack X®: Carboxen 569®	Thermal desorption	GC/MSD	Comparative study of the use of a Tenax TA® simple trap versus a trap formed by three carbon derived adsorbents for the VOC sampling in air	[135] (2010)
Tenax GR®	Thermal desorption	GC/MSD	Comparison of the VOC emissions produced by wood of different tree species subjected to different treatments	[233] (2010)
Tenax GR®	Thermal desorption	GC/MSD	Assignment of the chromatographic peaks in multicapillary column-ion mobility spectrometry by a comparative study versus TD/GC/MSD for breath analysis	[234] (2010)
Activated carbon	CH_2Cl_2	GC/MSD	Study of the VOC emissions released by particleboards produced in different conditions	[235] (2010)
Carboxen 100®: Carbopack B®	Thermal desorption	GC/MSD	Study of the monoaromatic and halogenated hydrocarbons analyzed in the gas phase and in the particulated matter in urban aerosols	[147] (2010)
Tenax TA®	Thermal desorption	GC/MSD	Study of the composition and temporal distribution of the sources of 51 VOCs in an Atlantic area	[236] (2010)
Carbotrap C®: Carbotrap B®: Carbosieve SIII®	Thermal desorption	GC/MSD	Study of the effect of the instantaneous and historic temperature on the BVOC emissions produced by *Pinus halepensis* and *Quercus ilex*. A leaf-cuvette was used as the enclosure system	[237] (2011)

Sampling(1)	Extraction	Ulterior analysis(2)	Application(3)	Citation (year)
Tenax TA®: Carbotrap®	Thermal desorption	GC/MSD	Quantification of the interferences reported with PTR-MS technique by comparison with thermal desorption/GC/MSD	[238] (2011)
Tenax TA®: Carbotrap®	Thermal desorption	GC/MSD or GC/FID	Study of the seasonal isoprene emission cycles produced by five oak tree species growing in South Korea. An enclosure system was used for the VOC sampling	[101] (2011)
Tenax TA®	Thermal desorption	GC/MSD	Determination of VOCs in indoor air sampled in buildings with low fresh-air ventilation rates placed in Bangkok	[142] (2011)
Chromosorb 106® (Passive sampling)	Thermal desorption	GC/FID	Study of the emission trends of VOCs in an industrial area placed in Kocaeli (Turkey)	[239] (2011)
Carbopack X®: Tenax® (Passive sampling)	Thermal desorption	GC/MSD	Study of the levels of exposure to VOCs in different microenvironments in Tianjin (China)	[240] (2011)

(1) Unless otherwise indicated, the sampling stages were developed forcing the entrance of air towards the sampling device (active sampling)

(2) GC: Gas Chromatography PID: PhotoIonization Detector FID: Flame Ionization Detector MSD: Mass Spectrometry Detector ECD: Electron Capture Detector GCxGC: Comprehensive Gas Chromatography TOF: Time Of Flight ELCD: Electrolytic Conductivity Detector

(3) BTEX: Benzene, Toluene, Ethylbenzene and Xylenes BTX: Benzene, Toluene and Xylenes NMHC: Non-Methane HydroCarbons TD/GC/MSD: Thermal Desorption/ Gas Chromatography/ Mass Spectrometry Detection GCxGC: Comprehensive Gas Chromatography PTR-MS: Proton Transfer Reaction- Mass Spectrometry

2.4. Other Systems for VOC Sampling

Although the methods summarized in the previous sections are the ones most frequently used for the VOC sampling, there are other methodologies whose applications are also referred in the literature.

Solid Phase Microextraction (SPME) is a technique frequently used in the analysis of VOCs in aqueous and solid matrices, although its application for the sampling of VOCs in environmental air samples is more restricted. This technique is based on the adsorption, absortion and desorption of the target compounds on polymeric fibres brought into contact with the sample or with its headspace. Its use in the analysis of environmental atmospheric air is limited due to the dilution showed by the VOCs in this matrix. This makes it difficult to achieve satisfactory detection limits. However, in the literature, there are some applications related to the sampling of VOCs in interior and confined atmospheres. For example, Lai and cols. [241] used this technique with fibres of PolyDiMethylSiloxane (PDMS) for the sampling of BVOCs considered as markers of marihuana, cocaine and methylenedioxy methylamphetamine with the objective of determining the camouflaged presence of such drugs in containers for transport. Martin and cols. [242] applied a system based on the use of fibres of PDMS for the determination of VOCs in exhaled breath vapour obtaining results faster and with lower detection limits than the ones achieved by the traditional methods.

In other application, Prikryl and Sevcik [243] characterized the interactions responsible for the sorption of different VOCs in air considering two individual kinds of fibres and other four combinations of several different fibres.

Derived from the SPME principles, in the recent years a series of techniques has been developed in order to get a major miniaturization of the sampling and concentration devices. Some of these techniques have been applied for the determination of VOCs, for example the Single Drop MicroExtraction (SDME), the Liquid-Phase MicroExtraction (LPME) or the Stir Bar Sorptive Extraction (SBSE) although these techniques find more application in the field of water analysis than in the analysis of air.

Other of technique derived from the SPME used for the air VOC sampling is the Needle Trap (NT) system. In these devices the fibres for SPME, or even conventional sorbents, are packed in a needle that is subsequently desorbed in an injector of a gas chromatograph. This kind of devices has been used by several authors [244-248].

The use of SemiPermeable Membrane Devices (SPMDs) for the VOC sampling in air is not either very much extended, since this technique finds its principal application in the analysis of water samples. However, Esteve-Turrillas and cols. [249] used semipermeable membranes for the sampling of BTEX (benzene, toluene, ethylbenzene, xylenes) in the interior air of vehicles and in petrol stations. These authors also used devices based on the utilization of semipermeable membranes for the sampling of VOCs in closed parkings and in diesel combustible boilers [128] or in the ambient air of different industries [129]. The system employed by these authors consists of a flat tube of low density polyethylene with a thickness lower than 100 μm which is permeable to the VOCs. The tube is filled with a substance (in this application trioleine) used to concentrate the target VOCs on its surface.

Liu and cols. [102] have used a portable Membrane Extraction with Sorbent Interface (MESI) system coupled to a Gas Chromatograph (MESI-GC) for the monitoring of the monoterpene emissions produced by the tree species *Eucalyptus dunnii*. Morley [250] also used a MESI-GC for the determination of ethylene in human breath. In this system the sampling and the concentration of the analytes is carried out in the same step.

Generically, the MESI instrumentation consists of a module in which a permeable membrane is located and one side of the membrane is exposed to the sample whereas the other side is in contact with the carrier gas. After leaving the membrane module, the carrier gas drags the analytes towards a microtrap filled with a proper adsorbent as, for example, Tenax TA®. This trap is kept at low temperature during the concentration stage and, after this concentration stage, the microtrap is heated in order to produce the thermal desorption of the analytes that are driven to the gas chromatograph.

3. EXTRACTION METHODS

The sampling with adsorbent solids is one of the most used methods for the analysis of VOCs in air. For that reason in the analytical procedure it is necessary to include a stage to extract the target compounds retained on the surface of the adsorbent bed. In this section the principal techniques used for the VOC extraction are summarized.

3.1. Solvent Extraction

The utilization of solvents for the extraction of VOCs retained on the surface of adsorbent solids is not a widely extended methodology for the analysis of environmnental samples. In fact, this method is not very recommended for this kind of applications since, generally, the VOC concentrations in the environmental matrices are very low. Thus, the extraction by solvents favours even more the dilution processes.

The choice of an appropriate solvent is restricted by the nature of the adsorbent used for the sampling but also by the characteristics of the target VOCs.

Among the advantages of this technique it is remarkable the possibility of repeating or making several analyses of the same sample. Besides this, solvent extraction is a relatively economic technique.

Among the inconveniences of the solvent extraction the most important one is the decrease of the sensitivity depending on the volume of solvent used and on the size of the aliquote analyzed. Other inconvenience to take into consideration is the possible introduction of interferences produced by the solvent. These interferences may affect the chromatographic separation of the target analytes, either as a consequence of the overlapping of the solvent peak with the peaks of the analytes or as a consequence of the release of monomers or short oligomer chains from the polymeric adsorbent [251]. Besides this, the use of solvents is accompanied by the problems derived from their toxicity [111]. Other inconvenience associated with this technique is that after the extraction process it is necessary to concentrate the extract, and the analytical recoveries will depend on the effectiveness of the concentration stage and on the kind of solvent used [252].

Nowadays, the use of solvents as extraction method is practically restricted to the VOC sampling in interior atmospheres. In this kind of applications it is very frequent to use activated carbon as the adsorbent and carbon disulfide as the extractant solvent, using mechanical stirring to improve the transfer process [253]. Campos-Candel and cols. [254] propose the desorption of the BTEX fraction retained on activated carbon by a Pressurised Liquid Extraction (PLE) using acetonitrile as extractant agent, although previously Richter [255] had already used the same technique for the direct extraction of the BTEX fraction charged on a solid matrix (sand) using dichloromethane as solvent. The Accelerated Solvent Extraction (ASE®) is a trademark from Dionex for the technique of PLE. In the ASE® the sample is confined in a metallic cell with the extraction solvent and this cell is subjected to high temperatures (50-200°C) and high pressures (500-3000 psi) for a short period of time (5-10 minutes). The combination of the use of solvents and high pressures and temperatures allows obtaining a more effective extraction of the analytes retained on the solid support. This technique is not compatible with the use of carbon disulfide as solvent for the extraction, since the autoignition point of this solvent is relatively low (approximately 100°C, whereas acetonitrile has an autoignition point of 500°C and the methylene acetate and methanol ones are about 400°C).

Besides the extraction by means of accelerated solvents the solvent ultrasonic extraction can be also carried out but in that case it is necessary to refrigerate the system to avoid the losses of the volatile compounds [256].

3.2. Thermal Desorption

Thermal desorption is a technique that allows the extraction of the volatile components of a non-volatile matrix by means of the heating of a sample/matrix in a flow of inert gas. The extracted volatile organic compounds can be optionally concentrated on a secondary adsorbent trap to be subsequently transferred into the chromatographic system.

This technique is compatible with solid, liquid and gaseous samples. In the case of solid and liquid samples the matrix can be subjected to a direct desorption whereas for liquid samples it is most frequent to use purge and trap systems to drag the analytes from the liquid sample and to concentrate them on an adsorbent trap.

To apply the thermal desorption technique to gaseous samples it is necessary to trap previously the volatile compounds on a solid adsorbent, as it was explained before.

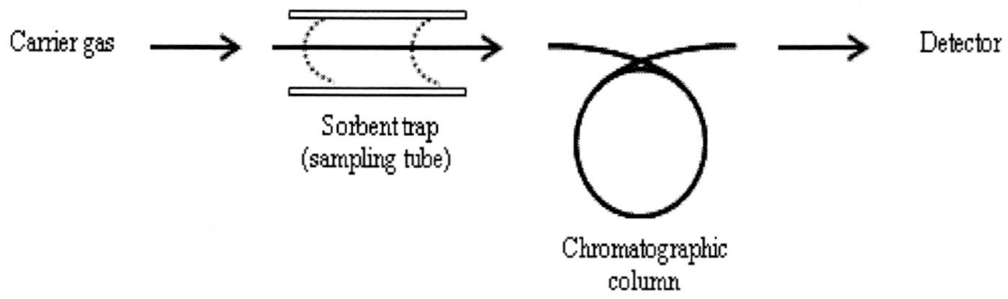

Figure 2. Thermal desorption process in a single stage.

The thermal desorption can be developed in one or in two stages. In the process of desorption in a single stage (Figure 2) the adsorbent solid is heated and the released volatiles are transferred directly to the chromatographic system. Before the extraction process, the adsorbent trap is purged under a flow of an inert gas, usually at ambient temperature, in order to eliminate the water vapour and the CO_2 retained on the adsorbent trap.

Although this method is quite simple, the application of thermal desorption in one stage produces a very pronounced broadening of the chromatographic bands and this makes the technique incompatible with the use of capillary columns in the chromatographic system.

This limitation can be solved subjecting the volatile compounds to a process of thermal desorption in two stages (Figure 3). In these procedures the trap used for the sampling is thermally desorbed, as it was explained previously, and the desorbed analytes are subsequently retained on a secondary trap by means of a new stage of cryoconcentration using temperatures in the range from -150°C to -30°C. This cryogenic trap may be filled with a small quantity of an appropriate adsorbent solid or glass beads, depending on the nature of the target analytes. After the cryoconcentration, the trap is quickly heated and the analytes are transferred towards the chromatographic column. The advantage of using this system is that it cryofocuses the sample, and this means that the second adsorption/desorption stage concentrates the sample on a small volume of adsorbent and it avoids the broadening of the chromatographic bands corresponding to the different analytes. This procedure improves the efficiency of the chromatographic separation and it is compatible with the use of capillary chromatographic columns.

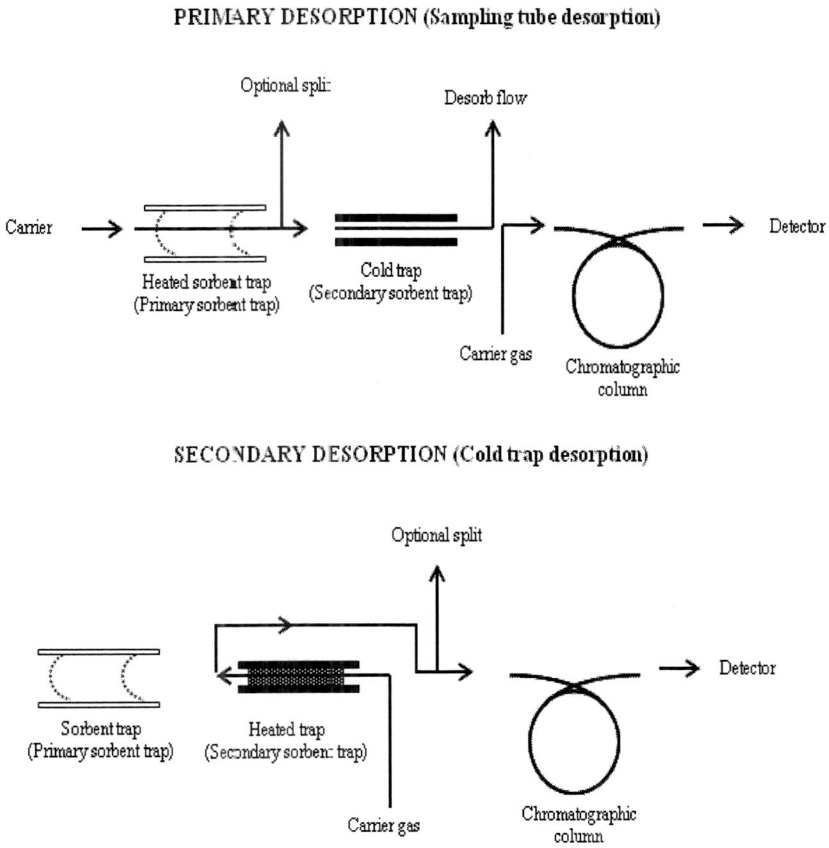

Figure 3. Thermal desorption process in two stages.

The techniques of thermal desorption have several advantages in comparison with the solvent extraction. Firstly, as in the thermal desorption techniques it is not produced a dilution of the samples, the detection limits achieved are better than the ones obtained using solvent extraction. This technique also provides better reproducibilities and major recovery percentages than the ones obtained by means of the solvent extraction [257].

Beside this, in the thermal desorption techniques the samples are not manipulated and the absence of solvents avoids the introduction of impurities derived from their use and also the generation of toxic residues. On the other hand, the adsorbent tubes can be reused several times after a relatively fast cleaning stage at high temperature before the sampling. Additionally the use of a specific adsorbent in the primary and secondary traps confers certain selectivity to the analysis. Finally, it is remarkable that the thermal desorption techniques are easier to automatize than the extraction methods based on the use of solvents.

Among the main inconveniences showed by the thermal desorption techniques it is outstanding that the possibility of analyzing the same sample for a second time is very restricted and it depends on the configuration of the instrumentation. Besides this, thermal desorption techniques are not compatible with thermolabile or with very heavy compounds which cannot be completely desorbed. Additionally, some compounds can suffer isomerization or dimerization reactions as a consequence of the application of high temperatures in the sequential thermal desorption stages. Apart from this, it has to be considered the high cost the analytical instrumentation. The price of this instrumentation is increased when the cooling in the secondary sorption/desorption stage is carried out using a cryogenic liquid. In this last case, the system will also occupy a big volume in the laboratory.

Nowadays there are commercially available some devices that couple a special injector in the chromatographic system to allow the cryoconcentration and focusing of the sample previously to the entrance in the chromatographic oven. Such devices are frequently used in combination with the canister methodology for the VOC sampling and they form a very compact thermal desorption unit. One of these commercially available systems is the one called Chromopack Thermal Desorption-Cold Trap Injection (TCT) from Chromopack.

The major inconvenience of these systems is that they require the use of a cryogenic liquid to carry out the cryofocusing of the sample and, as it was mentioned before, the use of these cryogenic liquids is expensive and requires a great volume in the laboratory.

4. Chromatographic Separation

Usually VOCs appear in the atmosphere as complex mixtures that have to be resolved by means of an appropriate technique. In this sense, capillary Gas Chromatography (GC) has been demonstrated to be a suitable technique to separate the individual compounds that integrate such complex mixtures and it is the most used technique for this purpose [258].

The separation is achieved using proper chromatographic columns with variable dimensions. Usually, the inner diameters vary between 0.20 and 0.53 mm and the length of the columns is in the range of 25 and 100 metres. The composition and polarity of the stationary phase depends on the chemical nature of the compounds that integrate the sample and this stationary phase can occupy different thickness varying from 1 to 5 μm, however in the analysis of semivolatile organic compounds the thickness of the stationary phase is usually lower than 1μm [259].

Traditionally, the chromatographic columns more frequently used for the analysis of VOCs are Wall Coated Open Tubular (WCOT) columns with silica gel as the stationary phase and Porous Layer Open Tubular (PLOT) columns with alumina as the stationary phase. These columns are commercialized under different names as for example AT-1, EC-1, CP-Sil5CB, DB-1, BP-1, HP-1, OV-1, RSL-150, RTX-1. Besides this, there is a group of columns specifically designed for the separation of VOCs according to the EPA methods (VOCOL, RTX-VMS, TRX-VGC, RTX-VRX or DB-VRX). These columns are designed using computer programs to achieve better and faster separations.

In the determination of biogenic VOCs, especially in the analysis of terpenoid VOCs, it is relatively frequent to use of columns with stationary phases based on cyclodextrins in order to separate the isomeric compounds [260].

Apart from the conventional stationary phases used in the chromatographic separations, nowadays there is an increasingly interest in the development of new stationary phases to obtain better resolutions in the separations. For example, Qi and Armstrong [261] used individual and combined monocationic and dicationic ionic liquids and combinations of ionic liquids with polydimethylsiloxane as stationary phases of columns of gas chromatography to achieve the separation of the volatile components of vegetal extracts. Ionic liquids have some interesting characteristics that make them very suitable for their use as stationary phases of columns of gas chromatography because they have low vapour pressures, high thermal stabilities and a dual nature that allows the separations of both polar and apolar compounds. However, these compounds also present a set of inconveniences to be used as stationary phases for GC, for example they do not form stable layers easily and they also show low thermal stability for some GC applications. Apart from this, they have a high retention and selectivity for specific compounds and low ability to produce chromatographic peaks with a Gaussian form. Among the ionic liquids, the dicationic ones show higher thermal stabilities and liquid ranges [261].

From the beginning of its development in the early years of the XX century, gas chromatography has experimented important achievements, not only referred to the development of new stationary phases but also to the implementation of new methodologies to resolve complex mixtures of compounds. The most conventional chromatographic technique consists in the use of a single column connected to a single detector (monodimensional chromatography). However, nowadays the use of MultiDimensional Gas Chromatography (MDGC) is reaching great popularity. In this technique discrete fractions of a sample, or even the complete sample, are subjected to two or more independent chromatographic separations. The MDGC started in the fifties and it has reached a consolidated position as a technique to increase the power of separation in several chromatographic applications. The MDGC can be divided into two great subclasses depending on the way in which the chromatographic development is carried out. On the one hand it is the conventional multidimensional gas chromatography, also called "heart-cutting MDGC", and on the other it is the more recently developed absolute MDGC, or "comprehensive MDGC" abbreviated as GCxGC.

In the heart-cutting gas chromatography one or more unresolved fractions eluting from the first column (or first dimension) are transferred to the second column (second dimension) which has a different polarity. The second dimension separates those compounds that the first column was unable to resolve [262, 263].

Usually, the first column has a non-polar stationary phase and a length between 30 and 60 metres, whereas the second of the columns is generally shorter (about 30 metres) and shows a higher polarity. Both columns can be placed in the same or in different ovens depending on the requirements of the analytical method.

As the first column has a non-polar nature, the separation is based on the differences of the boiling points of the analytes. Each fraction generated in the process of modulation is transferred into the second column and the compounds that constitute each fraction are quickly separated in this second dimension. Thus, those compounds that have a similar volatility are separated in the second column as a consequence of interactions based on the differences of polarity (for example, dipole-dipole, hydrogen bonds or polarization derived effects). The combination of two columns non-polar-polar is known as orthogonal configuration [264].

Each one of the unresolved fractions eluting from the first column can be directly transferred to the second column or can be retained on a cryogenic trap to be subsequently transferred to the second column. This allows carrying out a enrichment on the cryogenic trap with several fractions belonging to the same analyte that come from sequential injections.

In the year 1991, Liu and Phillips [265] introduced the technique denominated absolute multidimensional gas chromatography (GCxGC). With this technique all the components of a unique sample can be potentially subjected to a separation in two chromatographic columns working in series. The total time of the analysis is conditioned by the time elapsed in the first dimension. The columns used in the GCxGC are smaller than those used in the conventional MDGC. The typical lengths oscillate between 15 and 30 meters in the first column and only one or two in the second one. Thus, the length of the second column is small enough to achieve very fast separations as the fractions coming out from the first column are collected. The separation in the second column generally lasts from one to five seconds. This implies that this second column has not only a small length but also a reduced diameter (~0.1 mm) and also a stationary phase thickness very small (~0.1 µm). The use of two chromatographic columns to carry out the separation based on two different processes (volatility and polarity) allows that the compounds in the chromatograms appear organized as bands or groups of isomers easily identifiable [266].

The most important component in the instrumentation of GCxGC is the called modulator, which is placed at the head of the second column. The functions of this component are to isolate, reconcentrate and introduce continuously the small fractions eluting from the first column in the second one during the development of the chromatographic programme [264]. There are several kinds of modulators but, in general, these systems combine the cryofocusing with the thermal desorption of the trapped analytes. Initially, Liu and Phillips (1991) [265] used a thermal modulator in two stages that was heated by means of an electric resistance. This resistance was a painted film on a capillary surface and it was cooled by the action of the air. Among the main disadvantages of this modulator the most outstanding ones were the difficulty of its handling and its short lifetime, although it was used in some complex applications as for example in the determination of VOCs in kerosene fuels [267]. After that, it was developed a new thermal modulator using a wire instead of a painted film. This new modulator showed longer lifetimes but the linear response was more limited [268]. Phillips and cols. [269] designed a more robust modulator composed of a rotary oven that was displaced scanning a small fragment of the chromatographic column, producing pulses that were compatible with the chromatographic systems of GCxGC. Previously, Kinghorn and Marriott [270, 271] had developed a modulator that was not based on a heating phenomenon but on a cryogenic trapping. The working mode of this modulator consisted in the trapping and the sequentially releasing of the target analytes eluting from the first column by the movement of the cryogenic device along the second one. The main drawback of this device was that the water vapour condensed on the modulator surface producing ice. In order to solve these problems, other modulators based on the same principle but with different designs were subsequently developed [272, 273].

These modulator systems based on the scanning using high and low temperatures show the common inconvenience of causing the breaking of the extremes of the capillary chromatographic column.

One improvement over these modulators are the modulators based on the jet cooling or jet cooling/heating devices [273, 274] which do not have any mobile parts. The jet cooling/heating modulator uses two cold flows of nitrogen and another two heated flows that are applied as pulses to cool and heat alternatively two points at the beginning of the second column. This procedure focuses and mobilizes the analytes eluting from the first column in the second one. The main drawback of such systems is that they require enormous containers to store the cryogenic liquids used in the modulation process.

Apart from the thermal modulators, there is other category of modulators based in the use of different mechanical valves. This is the kind of modulators used in the heart-cutting MDGC.

The applications of the multidimensional gas chromatography (both comprehensive and heart-cutting modalities) for the analysis of VOCs in different atmospheric matrices are very abundant in the literature ([275] and references therein). In fact, there are several bibliographic reviews in which the fundaments and applications of such techniques are compiled [264, 276-281].

As a consequence of the enormous resolution power of GCxGC this technique has been applied for the analysis of samples as complex as kerosenes and fuels [266, 282], however in the literature there are some applications related to the analysis of atmospheric and aqueous matrices [191, 274, 283-285]. This technique has also found a wide application in the analysis of VOCs present in other kinds of matrices as, for example, essential oils, wines, etc [286, 287].

One of the main advantages of the multidimensional gas chromatography is that it has a higher power of resolution in comparison with the conventional chromatography. Besides this, multidimensional chromatography allows the analysis of compounds with a broad range of volatilities in a same sample. Additionally, the use of GCxGC produces a very important increase of the sensitivity of the analytical methods due to the narrowing of the chromatographic bands. It has been estimated that such increase in the sensitivity oscillates between a 25 and a 50% regarding the monodimensional chromatography. This range is quite broad since the kind of modulator determines the resolution achieved during the chromatographic process. Apart from this, the characteristics of the second column (length, inner diameter, thickness and characteristics of the stationary phase), the flow of carrier gas and the temperature programme also affect the sensitivity of the separation process.

Among the main limitations of the multidimensional gas chromatography it is outstanding the high cost of the instrumentation and also, in the case of GCxGC, the need of using a very fast detector and a powerful software in order to process such quantity of information. Additionally, the interpretation of the GCxGC chromatograms is quite difficult.

Besides the systems of multidimensional gas chromatography where the columns are placed in series, in other chromatographic configurations (monodimensional) the columns are collocated in a parallel way to achieve the VOC separation. For example, Jobson and cols. [288] describe a system for the VOC analysis based on the canister methodology for the sampling and on the separation using four chromatographs working in parallel. In that application each gas chromatograph is equipped with a different column: a PLOT Al_2O_3/Na_2SO_4 column for the analysis of compounds C_2-C_5, a DB-1 column for C_3-C_{10}, a Cyclodex-B for the compounds C_6-C_{1} and a DB5-MS column for the separation of the C_4-C_{11} compounds. Helmig and cols. [83] also used different columns and detector combinations working in parallel to study the VOCs and other compounds released from soils covered with snow.

Regarding the quantification methods, in the analysis of VOCs in atmospheric samples it is recommended to use internal standards to achieve good precisions and repetibilities [289].

5. Detection Systems

Regarding the detection systems used for the determination of VOCs in air samples the ones more frequently used are the Flame Ionization Detector (FID), the Mass Spectrometry Detector (MSD), the Electron Capture Detector (ECD) and the PhotoIonization Detector (PID). However other systems are also used as the ozone chemiluminescence detector, the isotopic relation mass spectrometry detector, the ionic mobility spectrometry detector or the electrolytic conductivity detector. All these systems will be briefly described in the following sections, besides this, in the Tables 1 and 3 they are shown some applications in which these kinds of detectors were used.

5.1. Flame Ionization Detector (FID)

The Flame Ionization Detector (FID) is a selective system for compounds with a hydrocarbonaceous nature, and it produces a signal for most of the VOCs. In this system, the compounds eluting from the chromatographic column with the carrier gas are mixed with hydrogen and they are burnt in air to produce a very hot flame that ionizes the hydrocarbons. A couple of electrodes, charged with a polarizing potential, recovers the ions and provides an electric current proportional to the number of generated ions [258]. This current is amplified and subsequently recorded.

This kind of detector has been used in the analysis of VOCs with an anthropogenic and a biogenic origin since it is sensitive to both classes of hydrocarbons showing a broad linear range. Additionally, FID allows the prediction of response factors for hydrocarbons belonging to a homologous series.

In the Tables 1 and 3 some applications in which the detection was carried out by means of flame ionization detectors are shown. Some authors have combined the use of flame ionization and capture electron detectors in series achieving better sensitivities than using these detectors working in a parallel way [162].

5.2. Mass Spectrometry Detector (MSD)

Mass Spectrometry (MS) has been widely used as a detection system for GC since the coupling GC-MSD is very useful to carry out the analysis of samples with an unknown composition [290]. A mass spectrometer consists of an ionization source, a mass filter or ion analyzer, a system to collect the formed ions and devices for the detection and amplification of the received signal. Besides this, it is also necessary the utilization of pumps to generate and keep the detection system under elevated vacuum conditions.

The most used method for obtaining ions in the detection of VOCs by mass spectrometry is the electronic ionization (EI+). In the EI+ technique the molecules of the sample are introduced in the detection system in a gaseous state and they are bombarded with a flow of electrons coming from a filament made of rhenium or tungsten (ionization source). In order to avoid the recombination of the ions and to get a path without collisions, it is necessary that the system is subjected to vacuum conditions.

As a consequence of the effect produced by the electronic current generated by the heated filament, the molecules of the compounds belonging to the sample are ionized positively and, after that, they are repelled by a small plate that is subjected to positive potentials. Thus, the generated ions are driven to a mass analyzer where they are electrically or magnetically separated depending on their mass/charge ratios. Finally, the ion current is detected and amplified.

Other method to produce ions from the gaseous sample is the chemical ionization (CI) method in which the gaseous flow containing the target analytes is not ionized directly by a current of electrons, but by ions coming from the previous electronic ionization of a reactive gas, usually methane.

The analysis of VOCs with biogenic origins is frequently developed by GC/MSD since the structures of such compounds, especially the structures of the monoterpenes, are usually very similar and in many occasions the only difference among them is the presence of double bonds or different functional groups in their structures.

Mass spectrometry coupled to GC systems offers the additional advantage of allowing the identification of the compounds using both the retention times and the information provided by the mass spectra.

The ion analyzers more frequently used for the analysis of VOCs in air are the quadrupoles since these analyzers provide high sensitivity in the separation of light ions. Ionic trap analyzers are used to a lesser extent because in these systems it is possible that the recombination of the formed ions takes place during the time that the ions last in the trap, producing losses of sensitivity.

In those applications in which the separation is achieved by means of GCxGC techniques, it is necessary to use very fast detection systems and, in this sense, the Mass Spectrometry Detection with an analyzer of Time Of Flight (MSD-TOF) is very useful.

Some authors use two systems of separation and quantification combined. In the case of the analysis of VOCs in air the most used combination is TD/GC/FID and TD/GC/MSD [95, 169, 291-295]. In some occasions the use of a second system of separation and detection is only to confirm the results obtained by other methodology but it is also frequent that the MSD system is used to identify the target analytes while the FID response is employed with quantification purposes [296].

5.3. Photoionization Detector (PID)

In the PhotoIonization Detector (PID) the molecules of the analytes that form the sample in gaseous state and with ionization potentials lower than the energy of a photon are ionized by means of the absorption of radiation emitted by a lamp in the far UV [258]. The ions generated by the UV radiation are driven from an acceleration electrode to a collector electrode where the formed current is measured. This is a practically non-destructive detection system and it is highly sensitive for most of the organic compounds, especially for the alkenes and the aromatic ones. In fact, the detection limits achieved with this detector are among a 10 and a 50% better than the ones obtained by the FID, as a consequence of the higher responses and of the lower background signals.

Song and cols. [297] improved the design of the standard photoionization detector modifying the geometry of the collector electrode and integrating a thermal preconcentration system in the PID. Using this system, these authors achieved a better stabilization of the signal and an increase of the sensitivity among 8 and 9 times regarding the conventional PID.

The PIDs are very common detection systems in the portable gas chromatographs (for example [100, 298, 299]).

5.4. Electron Capture Detector (ECD)

In the Electron Capture Detector (ECD) a radioactive isotope (for example ^{63}Ni) releases β particles that collide with molecules of a carrier gas producing a great quantity of electrons with low energy. These electrons are collected in electrodes generating a small background electronic current. If an analyte of the sample shows affinity for the electrons (for example, the halogenated compounds) it will capture a fraction of electrons from this current with low energy to form ions charged negatively. This will produce a fall down in the background signal that will be registered by the detector.

Besides the selectivity for the compounds with electronegative atoms, the ECD also allows reaching detection limits up to 10^4 better than the ones achieved using the FID detector. However, for the analysis of the molecules with two or less chlorine atoms the sensitivity is not as good. In order to solve this limitation the carrier gas can be doped with oxygen, since the presence of oxygenated ions in the system catalyzes the caption of electrons by the compounds of the sample. This detector with oxygen enrichment was used by Urhahn and Ballshmiter [300] for the determination of halogenated VOCs at trace concentrations in atmospheric air samples taken above the Atlantic Ocean.

5.5. Detection Using Ozone Chemiluminescence

This kind of detection is based on the fact that the reactions among the alkenes and ozone produce formaldehyde in an excited electronic state that later releases the excess of energy by the emission of light. The reactions that take place are:

$$\text{Alquenes} + O_3 \longrightarrow HCHO^* + \text{products}$$

$$HCHO^* \longrightarrow HCHO + h\nu$$

The emission produced by the excited formaldehyde takes place in the region between 450 and 550 nm and the intensity of the produced light can be measured and related to the concentration of alkenes in the sample.

The selectivity of the detector response is very sensitive to its temperature. Thus, at temperatures over 100°C only the alkenes can be detected while if the detector is over 250°C the alkanes also produce an analytical signal.

Based on this principle Hills and Zimmerman [301] built a detector for the on-line monitoring of isoprene, with response times of 10 seconds and detection limits of 400 pptv. Tambunan and cols. [103] used a very similar system for the determination of isoprene emitted by tropical tree species. Rasulov and cols. [302] also employed a portable chemiluminescence instrument for the determination of isoprene produced by poplars in absence of light.

5.6. Combustion/Isotope Ratio Mass Spectrometry

As the isotopic abundance of the carbon can differ depending on the origin (anthropogenic/ biogenic) of the hydrocarbons, the origin of a chemical compound can be established measuring the ratio $^{13}C/^{12}C$ for each compound. At the same time, the relative importance of the anthropogenic and biogenic sources can be evaluated for each analyte of the sample.

The coupling of this detection system to gas chromatography techniques produces the called GC-C-IRMS. (Gas Chromatography Combustion-Isotope Ratio Mass Spectrometry) techniques. The instrumentation consists of a gas chromatograph connected to a combustion furnace that acts as an interface, and an isotope ratio mass spectrometer. Thus, the analytes in a gaseous state are separated in the chromatographic system and, when the compounds elute from the column, they are converted into CO_2 by their combustion in a quartz tube that is electrically heated.

The isotope ratio mass spectrometry can be combined with the use of thermal desorption coupled to conventional gas chromatography. In fact Turner and cols. [303] have been pioneers in the specific isotopic analysis of different individual VOCs analyzed in atmospheric samples. Rudolph and cols. [304] proved that the study of the carbon isotope ratio for anthropogenic NMHC allows the quantitative estimation of the degradation processes by the OH radical whereas Rudolph and cols. [305] used carbon isotope ratios to study the atmospheric chemistry of isoprene.

5.7. Ion Mobility Spectrometry

Other detection system reported in the literature [234, 306, 307] is the Ion Mobility Spectrometry (IMS) combined with different sources of ionization. These sources can be classified as radioactive (for example ^{63}Ni) or non-radioactive (UltraViolet lamps, UV-IMS) sources. This technique has suffered a great demand for medical diagnosis, biological applications and process control [308].

There are two kinds of ion mobility detectors that are combined with the chromatographic separation. The first type is the one called DTIMS (Drif Tube Ion Mobility Spectrometry) with an instrumentation formed generically by a chamber of ionization equipped with a β-radiation or UV-radiation source, an ion-molecule injection shutter, an ion drift tube and a collector of ions. This technique is based on the identification and quantification of the analytes through their gas phase ion mobilities. Thus, the vaporized samples are ionized and the velocities of the produced ions are measured in a weak electric field in order to obtain the ion mobilities. This process allows the characterization of the vapour substances.

This technique is very fast and the required instrumentation is not very expensive; apart from that, these systems do not need vacuum conditions to operate and the analysis can be carried out at ambient temperature. For all these reasons this kind of detectors is frequently used in portable instruments for the on-site monitoring of VOCs.

A second type of the Ion Mobility Spectrometry Detector is the one called Differential Mobility Spectrometer (DMS). In this technique the ions are subjected alternately to strong and weak electric fields and the separation of the ions results from the mobility differences in both electric fields [309].

One of the most promising applications of IMS is its combination with MultiCapillary Columns (MCC-IMS) to achieve the separation of the target analytes using chromatographic columns formed by a set of capillary-columns working in parallel. These MCC are formed by more than 1000 capillaries, with standard sizes of 50-300 mm in length and approximately 3 mm for the total diameter of the set of capillaries. These capillaries are filled with a proper stationary phase [306].

One of the main drawbacks of this technique is that small variations in temperature and flow velocities produce changes in the retention times and this makes it difficult to identificate the target compounds [310].

5.8. Electrolytic Conductivity Detector

The ElectroLytic Conductivity Detector (ELCD) is mostly used for the determination of VOCs with a halogenated nature, although it is also useful for the analysis of compounds with nitrogen or sulphur atoms in their structures. The ELCD converts the halogenated compounds that elute from the GC column into an ionizable gas under reducting conditions in a high temperature catalytic reactor. Subsequently this gaseous current is driven into the cell of the reactor and it is solved into a deionized solvent producing an increase of the electrolytic conductivity of the sample. The detector amplifies the instantaneous change in the conductivity and generates a signal proportional to the quantity of halogenated atoms in the sample.

In the analysis of VOCs in air the use of this detector is sometimes combined with a PID working in series [168, 298] in order to get more information from the samples and not only information related to the halogenated compounds.

6. Other Systems of Analysis and Detection

Apart from the systems described previously in which it is necessary to incorporate a separation stage previous to the detection, there are other methods of analysis where a direct detection of the analytes is carried out.

Fourier Transform InfraRed (FTIR) spectroscopy is one of these methodologies. In this technique the infrared radiation is used to produce a variation in the vibrational states of the molecules of the sample. The absorption of radiation by the molecules in a specific interval of wavelengths indicates the kind of bonds and the functional groups of their structures. In this technique molecules with similar functional groups provide spectra slightly different and this is enough to identify the components of a sample. However, in multicomponent spectra the quantification of substances with a very similar structure and without functional groups is very complicated. This is a consequence of the difficulty of finding specific bands for each individual compound [311]. The use of FTIR spectroscopy is very frequent in the portable equipment.

Other of the techniques reported in the literature for VOC analysis in air is the Differential Optical Absorption Spectrometry (DOAS). This system has a lamp of emission, generally a lamp of xenon, whose light is received in a collector electrode and it is conducted towards a spectrometer using optical fibre. Inside the spectrometer the light is diffracted in its wavelengths in the range of the UV-VIS. Then, the refracted light is projected on a disk placed in front of a photomultiplier that amplifies the signal produced by the light that falls on the disk. Finally, the received signal is converted into a digital signal and it is registered by a computer. When a sample is introduced in the instrument, the analytes absorb light in the different wavelengths along the path that separates the emission source from the collector electrode. Thus, the photomultiplier measures the intensity of each wavelength through the optical path and the processor converts the signal into values of concentration [312].

The laser photoacoustic absorption spectroscopy is other of the methods used for the analysis of VOCs in atmospheric samples. The photoacoustic effect is based on the conversion of electromagnetic energy in an acoustic signal. The molecules of the analytes are excited by means of the absorption of light in the infrared range using a laser. After the absorption of energy, the molecules come back to their basal states by non-radiant processes (changes in their vibrational and rotational energies) and this produces an increase in the temperature and volume of the sample. When the absorption is modulated, the changes in the temperature and volume produce pressure (acoustic) oscillations that are detected using a very sensitive microphone.

The laser photoacoustic absorption spectroscopy was used by Zelinger and cols. [313] for the study and modeling of the pollutant dispersion in a tunnel and by Altuzar and cols. [69] in the monitoring of ethane in urban atmospheres. Besides this, Zelinger and cols. [314] demonstrated the ability of this technique to quantify compounds with very similar structures individually.

The last technique that will be treated in this section is the Proton Transfer Reaction Mass Spectrometry (PTR-MS) that nowadays has an important application in the study of BVOCs emitted by vegetation in real time and with a high sensitivity, [33, 100, 238, 315-319] although this technique has also been used in the study of anthropogenic VOCs [320-322]. PTR-MS is a kind of mass spectrometry based on the chemical ionization of the molecules using hydronium ions, H_3O^+, as the reactive ionic species. In this technique the H_3O^+ ion is generated *in situ* and it is mixed with the gaseous current of the sample that is introduced in the system. After this, mainly no-dissociative reactions take place. In that reactions it is produced the transfer of the proton to species with a major affinity towards the hydronium ion than the water, among these species most of the VOCs are included. The generated ions are driven to a mass spectrometer with a quadrupole as ion analyzer. The quadrupole separates the formed ions that subsequently will be quantified. A more detailed explanation of the PTR-MS operating principle can be found in the literature [323, 324]. Sometimes the mass spectrometer uses an ionic trap to separate the ions formed in the reaction with the H_3O^+ ion and in that case the technique receives the name of Proton transfer reaction Ion Trap Mass Spectrometry (PIT-MS) [89].

Some atmospheric components as O_2, N_2, CO_2 or light alkanes as methane, ethane or propane show a lower affinity towards the proton than the water, thus the transfer of the proton to these species does not take place and, consequently, they are not detected.

In this system there is not a previous separation of the formed ions generated from species with the same molecular mass, this is the case of the isomer compounds. For this reason the response for a specific mass is obtained as the sum of the ions coming from all the sources that produce an ion with this mass. The assignation of a specific species to the response obtained for a mass should be made carefully [113].

REFERENCES

[1] Cullis, C. F.; Hirschler, M. M. *Atmos. Environ.* 1989, *23*, 1195-1203.

[2] Guenther, A.; Hewitt, C. N.; Erickson, D.; Fall, R.; Geron, C.; Graedel, T.; Harley, P.; Klinger, L.; Lerdau, M.; Mckay, W. A.; Pierce, T.; Scholes, B.; Steinbrecher, R.; Tallamraju, R.; Taylor, J.; Zimmerman, P. *J. Geophys. Res. Atmospheres.* 1995, *100*, 8873-8892.

[3] Bocola, W.; Cirillo, M. C. *Atmos. Environ.* 1989, *23*, 17-24.

[4] Chagger, H. K.; Jones, J. M.; Pourkashanian, M.; Wiliams, A.; Owen, A.; Fynes, G. *Fuel.* 1999, *78*, 1527-1538.

[5] Fernández-Martínez, G.; López-Mahía, P.; Muniategui-Lorenzo, S.; Prada-Rodríguez, D.; Fernández-Fernández, E. *Atmos. Environ.* 2001, *35*, 5823-5831.

[6] Pulles, T.; Heslinga, D. *Atmos. Environ.* 2004, *38*, 3827-3838.

[7] Schauer, J. J.; Kleeman, M. J.; Cass, G. R.; Simoneit, B. R. T. *Environ. Sci. Technol.* 2001, *35*, 1716-1728.

[8] Wang, S. X.; Wei, W.; Du, L.; Li, G. H.; Hao, J. M. *Atmos. Environ.* 2009, *43*, 4148-4154.

[9] Cetin, E.; Odabasi, M.; Seyfioglu, R. *Sci. Total Environ.* 2003, *312*, 103-112.

[10] Chen, C. L.; Shu, C. M.; Fang, H. Y. *Environ. Monit. Assess.* 2006, *120*, 487-498.

[11] Jobson, B. T.; Berkowitz, C. M.; Kuster, W. C.; Goldan, P. D.; Williams, E. J.; Fesenfeld, F. C.; Apel, E. C.; Karl, T.; Lonneman, W. A.; Riemer, D. *J. Geophys. Res. Atmospheres.* 2004, *109*,

[12] Amin, T. A.; Bindra, S.; Narang, R. S. *J. AOAC Int.* 1998, *81*, 1027-1032.

[13] Niedojadlo, A.; Becker, K. H.; Kurtenbach, R.; Wiesen, P. *Atmos. Environ.* 2007, *41*, 7108-7126.

[14] Carr, D.; Von Ehrenstein, O.; Weiland, S.; Wagner, C.; Wellie, O.; Nicolai, T.; Von Mutius, E. *Environ. Res. Section A.* 2002, *90*, 111-118.

[15] John, C.; Friedrich, R.; Staehelin, J.; Schläpfer, K.; Stahel, W. A. *Atmos. Environ.* 1999, *33*, 3367-3376.

[16] Murena, F. *J. Environ. Sci. China.* 2007, *19*, 578-583.

[17] Belalcazar, L. C.; Fuhrer, O.; Ho, M. D.; Zarate, E.; Clappier, A. *Atmos. Environ.* 2009, *43*, 5830-5837.

[18] Srivastava, A.; Majumdar, D. *Environ. Monit. Assess.* 2010, *160*, 315-322.

[19] Baldasano, J. M.; Delgado, R.; Calbó, J. *Environ. Sci. Technol.* 1998, *32*, 405-412.

[20] Gros, V.; Williams, J.; Van Aardenne, J. A.; Salisbury, G.; Hofmann, R.; Lawrence, M. G.; Von Kuhlmann, R.; Lelieveld, J.; Krol, M.; Berresheim, H.; Lobert, J. M.; Atlas, E. *Atmos. Chem. Phys.* 2003, *3*, 1223-1235.
[21] Rudolph, J.; Von Czapiewski, K.; Koppmann, R. *Geophys. Res. Lett.* 2000, *27*, 1887-1890.
[22] Celebi, U. B.; Vardar, N. *Atmos. Environ.* 2008, *42*, 5685-5695.
[23] Atkinson, R. *Atmos. Environ.* 1990, *24 A*, 1-41.
[24] Atkinson, R.; Arey, J. *Chem. Rev.* 2003, *103*, 4605-4638.
[25] Czoschke, N. M.; Jang, M.; Kamens, R. M. *Atmos. Environ.* 2003, *37*, 4287-4299.
[26] Hu, D.; Kamens, R. M. *Atmos. Environ.* 2007, *41*, 6465-6477.
[27] Iinuma, Y.; Böge, O.; Gnauk, T.; Herrmann, H. *Atmos. Environ.* 2004, *38*, 761-773.
[28] Northcross, A. L.; Jang, M. *Atmos. Environ.* 2007, *41*, 1483-1493.
[29] Matsunaga, S.; Mochida, M.; Kawamura, K. *Atmos. Environ.* 2003, *37*, 2045-2050.
[30] Yu, S. *Atmos. Res.* 2000, *53*, 185-217.
[31] Wang, Y.; Kim, H.; Paulson, S. E. *Atmos. Environ.* 2011, *45*, 3149-3156.
[32] Hovorka, J.; Branis, M. *Atmos. Environ.* 2011, *45*, 2736-2749.
[33] Peñuelas, J.; Staudt, M. *Trends Plant Sci.* 2010, *15*, 133-144.
[34] Badol, C.; Borbon, A.; Locoge, N.; Léonardis, T.; Galloo, J.-C. *Anal. Bioanal. Chem.* 2004, *378*, 1815-1824.
[35] Tumbiolo, S.; Gal, J.-F.; Maria, P.-C.; Zerbinati, O. *Anal. Bioanal. Chem.* 2004, *380*, 824-830.
[36] Palluau, F.; Mirabel, P.; Millet, M. *Environ. Monit. Assess.* 2007, *127*, 177-187.
[37] Camel, V.; Caude, M. *J. Chromatogr. A* 1995, *710*, 3-19.
[38] Dettmer, K.; Engewald, W. *Anal. Bioanal. Chem.* 2002, *373*, 490-500.
[39] Król, S.; Zabiegala, B.; Namiesnik, J. *TrAC Trends Anal. Chem.* 2010, *29*, 1101-1112.
[40] Van Harreveld, A. P. *J. Air Waste Manag. Assoc.* 2003, *53*, 51-60.
[41] Wang, D. K. W.; Austin, C. C. *Anal. Bioanal. Chem.* 2006, *386*, 1099-1120.
[42] Bailey, J. C.; Schmidl, B.; Williams, M. L. *Atmos. Environ.* 1990, *24 A*, 43-52.
[43] Jordan, A.; Harnisch, J.; Borchers, R.; Le Guern, F.; Shinohara, H. *Environ. Sci. Technol.* 2000, *34*, 1122-1124.
[44] Liu, C.; Xu, Z.; Du, Y.; Guo, H. *Atmos. Environ.* 2000, *34*, 4459-4466.
[45] Montells, R.; Aceves, M.; Grimalt, O. *Environ. Monit. Assess.* 2000, *62*, 1-14.
[46] Pratt, G. C.; Palmer, K.; Wu, C. Y.; Oliaei, F.; Hollerbach, C.; Fenske, M. J. *Environ. Health Perspect.* 2000, *108*, 815-825.
[47] Colman, J. J.; Swanson, A. L.; Meinardi, S.; S., B. C.; Blake, D. R.; Rowland, F. S. *Anal. Chem.* 2001, *73*, 3723-3731.
[48] Na, K.; Kim, Y. P. *Atmos. Environ.* 2001, *35*, 2603-2614.
[49] Lee, S.; Li, W.-M.; Ao, C.-H. *Atmos. Environ.* 2002, *36*, 225-237.
[50] Hellén, H.; Hakola, H.; Laurila, T. *Atmos. Environ.* 2003, *37*, 1413-1424.
[51] Hsieh, C. C.; Tsai, J. H. *Chemosphere* 2003, *50*, 545-556.
[52] Kwok, N.-H.; Lee, S.-C.; Guo, H.; Hung, W.-T. *Build. Environ.* 2003, *38*, 1019-1026.
[53] Tsai, S. M.; Zhang, J.; Smith, K. R.; Ma, Y.; Rasmussen, R. A.; Khalil, M. A. K. *Environ. Sci. Technol.* 2003, *37*, 2869-2877.
[54] Kleinman, L. I.; Daum, P. H.; Lee, Y. N.; Nunnermacker, L. J.; Springston, S. R.; Weinstein-Lloyd, J.; Hyde, P.; Doskey, P.; Rudolph, J.; Fast, J.; Berkowitz, C. *J. Geophys. Res. Atmospheres.* 2003, *108*, 4096; 1-14.

[55] Lau, W.-L.; Chan, L.-Y. *Sci. Total Environ.* 2003, *308*, 143-155.
[56] Lee, S. C.; Kwok, N. H.; Guo, H.; Hung, W. T. *Sci. Total Environ.* 2003, *302*, 75-84.
[57] Na, K.; Pyo, K. Y.; Choo, K. M. *Atmos. Environ.* 2003, *37*, 733-742.
[58] Wang, H.-C.; Liu, H.-L. *Bull. Environ. Contam. Toxicol.* 2003, *71*, 856-859.
[59] Cox, M. L.; Fraser, P. J.; Sturrock, G. A.; Siems, S. T.; Porter, L. W. *Atmos. Environ.* 2004, *38*, 3839-3852.
[60] Gómez-Perales, J. E.; Colvile, R. N.; Nieuwenhuijsen, M. J.; Fernández-Bremauntz, A.; Gutiérrez-Avedoy, V. J.; Páramo-Figureoa, V. H.; Blanco-Jiménez, S.; Bueno-López, E.; Mandujano, F.; Bernabé-Cabanillas, R.; Ortiz Segovia, E. *Atmos. Environ.* 2004, *38*, 1219-1229.
[61] Guo, H.; Lee, S. C.; Chan, L. Y.; Li, W. M. *Environ. Res.* 2004, *94*, 57-66.
[62] Guo, H.; Wang, T.; Louie, P. K. K. *Environ. Pollut.* 2004, *129*, 489-498.
[63] Kato, S.; Kajii, Y.; Itokazu, R.; Hirokawa, J.; Koda, S.; Kinjo, Y. *Atmos. Environ.* 2004, *38*, 2975-2981.
[64] Pratt, G. C.; Yi Wu, C.; Bock, D.; Adgate, J. L.; Ramachandran, G.; Stock, T. H.; Morandi, M.; Sexton, K. *Environ. Sci. Technol.* 2004, *38*, 1949-1959.
[65] Lee, S. C.; Wang, B. *Atmos. Environ.* 2004, *38*, 941-951.
[66] Mcgaughey, G. R.; Desai, N. R.; Allen, D. T.; Seila, R. L.; Lonneman, W. A.; Fraser, M. P.; Harley, R. A.; Pollack, A. K.; Ivy, J. M.; Price, J. H. *Atmos. Environ.* 2004, *38*, 3363-3372.
[67] US. Environmental Protection Agency. *Compendium Method TO-15. Determination of volatile organic compounds (VOCs) in air collected in specially-prepared canisters and analyzed by gas chromatography/mass spectrometry*; US. EPA: Washington,DC, 1997.
[68] Aleksic, N.; Boynton, G.; Sistla, G.; Perry, J. *Atmos. Environ.* 2005, *39*, 7894-7905.
[69] Altuzar, V.; Tomás, S. A.; Zelaya-Angel, O.; Sánchez-Sinencio, F.; Arriaga, J. L. *Atmos. Environ.* 2005, *39*, 5219-5225.
[70] Barletta, B.; Meinardi, S.; Rowland, F. S.; Chan, C. Y.; Wang, X.; Zou, S.; Chan, L. Y.; Blake, D. R. *Atmos. Environ.* 2005, *39*, 5979-5990.
[71] Kim, J.-C.; Kim, K.-J.; Kim, D.-S.; Han, J.-S. *Chemosphere.* 2005, *59*, 1685-1696.
[72] Latella, A.; Stani, G.; Cobelli, L.; Duane, M.; Junninen, H.; Astorga, C.; Larsen, B. R. *J. Chromatogr. A* 2005, *1071*, 29-39.
[73] Anderson, B. E.; Chen, G.; Blake, D. R. *Atmos. Environ.* 2006, *40*, 3601-3612.
[74] Bai, J. H.; Baker, B.; Liang, B. S.; Greenberg, J.; Guenther, A. *Atmos. Environ.* 2006, *40*, 5753-5758.
[75] Barletta, B.; Meinardi, S.; Simpson, I. J.; Rowland, F. S.; Chan, C. Y.; Wang, X. M.; Zou, S. C.; Chan, L. Y.; Blake, D. R. *Atmos. Environ.* 2006, *40*, 7706-7719.
[76] Baker, A. K.; Beyersdorf, A. J.; Doezema, L. A.; Katzenstein, A.; Meinardi, S.; Simpson, I. J.; Blake, D. R.; Rowland, F. S. *Atmos. Environ.* 2008, *42*, 170-182.
[77] Barletta, B.; Meinardi, S.; Simpson, I. J.; Zou, S. C.; Rowland, F. S.; Blake, D. R. *Atmos. Environ.* 2008, *42*, 4393-4408.
[78] Kim, K. H.; Park, S. Y. *Atmos. Environ.* 2008, *42*, 5061-5070.
[79] Nelson, P. F.; Tibbett, A. R.; Day, S. J. *Atmos. Environ.* 2008, *42*, 5291-5303.
[80] Stocco, C.; Macneill, M.; Wang, D.; Xu, X. H.; Guay, M.; Brook, J.; Wheeler, A. J. *Atmos. Environ.* 2008, *42*, 5905-5912.
[81] Tang, J. H.; Chan, L. Y.; Chan, C. Y.; Li, Y. S.; Chang, C. C.; Wang, X. M.; Zou, S. C.; Barletta, B.; Blake, D. R.; Wu, D. *Atmos. Environ.* 2008, *42*, 3780-3794.

[82] Geng, F. H.; Cai, C. J.; Tie, X. X.; Yu, Q.; An, J. L.; Peng, L.; Zhou, G. Q.; Xu, J. M. *J. Atmos. Chem.* 2009, *62*, 229-247.
[83] Helmig, D.; Apel, E.; Blake, D.; Ganzeveld, L.; Lefer, B. L.; Meinardi, S.; Swanson, A. L. *Biogeochemistry.* 2009, *95*, 167-183.
[84] Mochalski, P.; Wzorek, B.; Sliwka, I.; Amann, A. *J. Chromatogr. B Analyt. Technol. Biomed. Life Sci.* 2009, *877*, 189-196.
[85] Mochalski, P.; Wzorek, B.; Sliwka, I.; Amann, A. *J. Chromatogr. B Analyt. Technol. Biomed. Life Sci.* 2009, *877*, 1856-1866.
[86] Beyersdorf, A. J.; Blake, D. R.; Swanson, A.; Meinardi, S.; Rowland, F. S.; Davis, D. *Atmos. Environ.* 2010, *44*, 4565-4574.
[87] Chung, M. Y.; Beene, M.; Ashkan, S.; Krauter, C.; Hasson, A. S. *Atmos. Environ.* 2010, *44*, 786-794.
[88] Matsunaga, S. N.; Chatani, S.; Morikawa, S.; Nakatsuka, S.; Suthawaree, J.; Tajima, Y.; Kato, S.; Kajii, Y.; Minoura, H. *Atmos. Environ.* 2010, *44*, 4982-4993.
[89] Bon, D. M.; Ulbrich, I. M.; De Gouw, J. A.; Warneke, C.; Kuster, W. C.; Alexander, M. L.; Baker, A.; Beyersdorf, A. J.; Blake, D.; Fall, R.; Jimenez, J. L.; Herndon, S. C.; Huey, L. G.; Knighton, W. B.; Ortega, J.; Springston, S.; Vargas, O. *Atmos. Chem. Phys.* 2011, *11*, 2399-2421.
[90] Guo, H.; Zou, S. C.; Tsai, W. Y.; Chan, L. Y.; Blake, D. R. *Atmos. Environ.* 2011, *45*, 2711-2721.
[91] Nichol, J.; Wong, M. S. *Atmos. Environ.* 2011, *45*, 2937-2943.
[92] Yao, H. Q.; Choi, H. L.; Zhu, K.; Lee, J. H. *Atmos. Environ.* 2011, *45*, 2577-2584.
[93] Cariou, S.; Guillot, J.-M. *Anal. Bioanal. Chem.* 2006, *384*, 468-474.
[94] Stump, F. D.; Dropkin, D. L. *Anal. Chem.* 1985, *57*, 2629-2634.
[95] Boissard, C.; Cao, X.-L.; Juan, C.-Y.; Hewitt, C. N.; Gallagher, M. *Atmos. Environ.* 2001, *35*, 917-927.
[96] Harrison, D.; Hunter, M. C.; Lewis, A. C.; Seakins, P. W.; Bonsang, B.; Gros, V.; Kanakidou, M.; Touaty, M.; Kavouras, I.; Mihalopoulos, N.; Stephanou, E.; Alves, C.; Nunes, T.; Pio, C. *Atmos. Environ.* 2001, *35*, 4699-4711.
[97] Sabillón, D.; Cremades, L. V. *Atmos. Environ.* 2001, *35*, 4419-4431.
[98] Simon, V.; Dumergues, L.; Bouchou, P.; Torres, L.; López, A. *Atmos. Res.* 2005, *74*, 49-63.
[99] Lin, C.; Owen, S. M.; Penuelas, J. *Soil Biol. Biochem.* 2007, *39*, 951-960.
[100] Geron, C.; Guenther, A.; Greenberg, J.; Karl, T.; Rasmussen, R. *Atmos. Environ.* 2006, *40*, 1645-1660.
[101] Lim, Y. J.; Armendariz, A.; Son, Y. S.; Kim, J. C. *Atmos. Environ.* 2011, *45*, 2202-2210.
[102] Liu, X.; Pawliszyn, R.; Wang, L.; Pawliszyn, J. *Analyst.* 2004, *129*, 55-62.
[103] Tambunan, P.; Baba, S.; Kuniyoshi, A.; Iwasaki, H.; Nakamura, T.; Yamasaki, H.; Oku, H. *Chemosphere.* 2006, *65*, 2138-2144.
[104] Rinnan, R.; Rinnan, Å.; Faubert, P.; Tiiva, P.; Holopainen, J. K.; Michelsen, A. *Environ. Exp. Bot.* doi: DOI: 10.1016/j.envexpbot.2010.11.006, *In Press, Corrected Proof,*
[105] US. Environmental Protection Agency. *Compendium method TO-14: The determination of volatile organic compounds (VOCs) in ambient air using summa passivated canister sampling and gas chromatographic analysis*; US. EPA: Washington, DC, 1988.

[106] US. Environmental Protection Agency. *Compendium Method TO-14 A. Determination of volatile organic compounds (VOCs) in ambient air using specially prepared canisters with subsequent analysis by gas chromatography*; US. EPA: Washington, DC, 1997.
[107] Kelly, T. J.; Holdren, M. W. *Atmos. Environ.* 1995, *29*, 2595-2608.
[108] Batterman, S. A.; Zhang, G.-H.; Baumann, M. *Atmos. Environ.* 1998, *32*, 1647-1655.
[109] Rhoderick, G. C. *Anal. Bioanal. Chem.* 2010, *398*, 1417-1425.
[110] Palluau, F.; Mirabel, P.; Millet, M. *Anal. Bioanal. Chem.* 2005, *381*, 742-761.
[111] Des Tombe, K.; Verma, D. K.; Stewart, L.; Reckzek, E. B. *AIHA Journal.* 1991, *52*, 136-144.
[112] Plass-Dulmer, C.; Michl, K.; Ruf, R.; Berresheim, H. *J. Chromatogr. A* 2002, *953*, 175-197.
[113] Kuster, W. C.; Jobson, B. T.; Karl, T.; Riemer, D.; Apel, E.; Goldan, P. D.; Fehsenfeld, F. C. *Environ. Sci. Technol.* 2004, *38*, 221-228.
[114] Derendorp, L.; Holzinger, R.; Wishkerman, A.; Keppler, F.; Röckmann, T. *Atmos. Environ.* 2011, *45*, 3112-3119.
[115] Wang, J. L.; Chen, W. L.; Lin, Y. H.; Tsai, C. H. *J. Chromatogr. A* 2000, *896*, 31-39.
[116] US. Environmental Protection Agency. *Compendium method TO-17: Determination of volatile organic compounds in ambient air using active sampling onto sorbent tubes*; US. EPA: Washington, DC, 1999.
[117] Brown, R. H.; Wright, M. D. *Analyst.* 1994, *119*, 75-77.
[118] Brown, R. H. *J. Environ. Monit.* 2000, *2*, 1-9.
[119] Cao, X.-L.; Hewitt, N. *Environ. Technol.* 1991, *12*, 1055-1062.
[120] Chung, C.-W.; Morandi, M. T.; Stock, T. H.; Afshar, M. *Environ. Sci. Technol.* 1999, *33*, 3666-3671.
[121] Roche, A.; Thevenet, R.; Jacob, V.; Kaluzny, P.; Ferrari, C.; Baussand, P.; Foster , P. *Atmos. Environ.* 1999, *33*, 1905-1912.
[122] Bomboi, M. T.; Pérez, A.; Rodriguez, B.; Galán, D.; Díaz, A.; Fernández, R. *Fresenius Environ. Bull.* 2002, *11*, 437-440.
[123] Sexton, K.; Adgate, J. L.; Mongin, S. J.; Pratt, G. C.; Ramachandran, G.; Stock, T. H.; Morandi, M. T. *Environ. Sci. Technol.* 2004, *38*, 2593-2602.
[124] Sexton, K.; Adgate, J. L.; Ramachandran, G.; Pratt, G. C.; Mongin, S. J.; Stock, T. H.; Morandi, M. T. *Environ. Sci. Technol.* 2004, *38*, 423-430.
[125] Begerow, J.; Jermann, E.; Keles, T.; Dunemann, L. *Fresen. J. Anal. Chem.* 1999, *363*, 399-403.
[126] Pennequin-Cardinal, A.; Plaisance, H.; Locoge, N.; Ramalho, O.; Kirchner, S.; Galloo, J.-C. *Talanta.* 2005, *65*, 1233-1240.
[127] Esteve-Turrillas, F. A.; Armenta, S.; Garrigues, S.; Pastor, A.; De La Guardia, M. *Anal. Chim. Acta.* 2007, *587*, 89-96.
[128] Esteve-Turrillas, F. A.; Ly-Verdu, S.; Pastor, A.; De La Guardia, M. *J. Chromatogr. A* 2009, *1216*, 8549-8556.
[129] Ly-Verdu, S.; Esteve-Turrillas, F. A.; Pastor, A.; De La Guardia, M. *Anal. Chim. Acta.* 2010, *677*, 131-139.
[130] Simon, V.; Riba, M. L.; Waldhart, A.; Torres, L. *J. Chromatogr. A* 1995, *704*, 465-471.
[131] Volden, J.; Thomassen, Y.; Greibrokk, T.; Thorud, S.; Molander, P. *Anal. Chim. Acta.* 2005, *530*, 263-271.

[132] Haunold, A.; Rosenberg, E.; Grasserbauer, M. *Int. J. Environ. Anal. Chem.* 1997, *67*, 157-172.

[133] Roukos, J.; Plaisance, H.; Leonardis, T.; Bates, M.; Locoge, N. *J. Chromatogr. A* 2009, *1216*, 8642-8651.

[134] Borusiewicz, R.; Zieba-Palus, J. *J. Forensic Sci.* 2007, *52*, 70-74.

[135] Gallego, E.; Roca, F. J.; Perales, J. F.; Guardino, X. *Talanta.* 2010, *81*, 916-924.

[136] Dettmer, K.; Knobloch, T.; Engewald, W. *Fresen. J. Anal. Chem.* 2000, *366*, 70-78.

[137] Cao, X.-L.; Hewitt, C. N. *J Chromatogr. A* 1994, *688*, 368-374.

[138] Pérez-Rial, D.; Peñuelas, J.; López-Mahia, P.; Llusià, J. *J. Environ. Monit.* 2009, *11*, 1268-1275.

[139] Walling, J. F.; Bumgarner, J. E.; Driscoll, D. J.; Morris, C. M.; Riley, A. E.; Wright, L. H. *Atmos. Environ.* 1986, *26*, 51-57.

[140] Bates, M.; González-Flesca, N.; Sokhi, R.; Cocheo, V. *Environ. Monit. Assess.* 2000, *65*, 89-97.

[141] Pommer, L.; Fick, J.; Andersson, B.; Nilsson, C. *Atmos. Environ.* 2002, *36*, 1443-1452.

[142] Ongwandee, M.; Moonrinta, R.; Panyametheekul, S.; Tangbanluekal, C.; Morrison, G. *Build. Environ.* 2011, *46*, 1512-1522.

[143] Isidorov, V. A.; Zenkevich, I. G.; Ioffe, B. V. *Atmos. Environ.* 1985, *19*, 1-8.

[144] Durana, N.; Navazo, M.; Alonso, L.; García, J. A.; Ilardia, J. L.; Gómez, M. C.; Gangoiti, G. *J. Air Waste Manag. Assoc.* 2002, *52*, 1176-1185.

[145] Navazo, M.; Durana, N.; Alonso, L.; Gómez, M. C.; García, J. A.; Ilardia, J. L.; Gangoiti, G.; Iza, J. *Environ. Monit. Assess.* 2008, *136*, 53-68.

[146] Seeley, I.; Broadway, G. *Fresenius Environ. Bull.* 1994, *3*, 158-163.

[147] Matsumoto, K.; Matsumoto, K.; Mizuno, R.; Igawa, M. *Atmos. Res.* 2010, *97*, 124-128.

[148] Durana, N.; Navazo, M.; Gomez, M. C.; Alonso, L.; Garcia, J. A.; Ilardia, J. L.; Gangoiti, G.; Iza, J. *Atmos Environ.* 2006, *40*, 2860-2872.

[149] Hong, D. W.; Heo, G. S.; Han, J. S.; Cho, S. Y. *Atmos. Environ.* 2004, *38*, 5567-5576.

[150] Harper, M. *Analyst.* 1994, *119*, 65-69.

[151] Brosillon, S.; Manero, M.-H.; Foussard, J.-N. *Environ. Sci. Technol.* 2001, *35*, 3571-3575.

[152] Matisová, E.; Skrabáková, S. *J. Chromatogr. A* 1995, *707*, 145-179.

[153] Karanfil, T.; Kilduff, J. E. *Environ. Sci. Technol.* 1999, *33*, 3217-3224.

[154] Delage, F.; Pré, P.; Le Cloirec, P. *Environ. Sci. Technol.* 2000, *34*, 4816-4821.

[155] Wang, D. K. W.; Austin, C. C. *Anal. Bioanal. Chem.* 2006, *386*, 1089-1098.

[156] Matisová, E.; Skrabáková, S. *Anal. Chim. Acta.* 1995, *309*, 181-188.

[157] Patil, S. F.; Lonkar, S. T. *J. Chromatogr. A* 1994, *684*, 133-142.

[158] Clark, A. I.; Mcintyre, A. E.; Perry, R.; Lester, J. N. *Environ. Pollut. Series B.* 1984, *7*, 141-158.

[159] Field, R. A.; Goldstone, M. E.; Lester, J. N.; Perry, R. *Environ. Technol.* 1994, *15*, 801-812.

[160] Karanassios, I. G.; Georgakiles, V. I.; Lahaniastis, E. S.; Pilidis, G. A. *Fresenius Environ. Bull.* 1994, *3*, 511-516.

[161] Lindskog, A.; Moldanová, J. *Atmos. Environ.* 1994, *28*, 2383-2398.

[162] Begerow, J.; Jermann, E.; Keles, T.; Koch, T.; Dunemann, L. *J. Chromatogr. A* 1996, *749*, 181-191.

[163] Bemgård, A.; Colmsjö, A.; Melin, J. *J. Chromatogr. A* 1996, *723*, 301-311.

[164] Röse, U. S. R.; Manukian, A.; Heath, R. R.; Tumlinson, J. H. *Plant. Physiol.* 1996, *111*, 487-495.
[165] Larsen, B.; Bomboi-Mingarro, T.; Brancaleoni, E.; Calogirou, A.; Cecinato, A.; Coeur, C.; Chatzianestis, I.; Duane, M.; Frattoni, M.; Fugit, J.-L.; Hansen, U.; Jacob, V.; Mimikos, N.; Hoffmann, T.; Owen, S.; Pérez-Pastor, R.; Reichmann, A.; Seufert, G.; Staudt, M.; Steinbrecher, R. *Atmos. Environ.* 1997, *31*, 35-49.
[166] Steinbrecher, R.; Hauff, K.; Rabong, R.; Steinbrecher, J. *Atmos. Environ.* 1997, *31*, 79-88.
[167] He, C.; Murray, F.; Lyons, T. *Atmos. Environ.* 2000, *34*, 645-655.
[168] Yamamoto, N.; Okayasu, H.; Murayama, S.; Mori, S.; Hunahashi, K.; Suzuki, K. *Atmos. Environ.* 2000, *34*, 4441-4446.
[169] Sartin, J. H.; Halsall, C. J.; Davison, B.; Owen, S.; Hewitt, C. N. *Anal. Chim. Acta.* 2001, *428*, 61-72.
[170] Binnie, J.; Cape, J. N.; Makie, N.; Leight, I. D. *Sci. Total Environ.* 2002, *285*, 53-67.
[171] Chen, T.-Y.; Li, M.-J.; Wang, J.-L. *J. Chromatogr. A* 2002, *976*, 39-45.
[172] Singer, B. C.; Hodgson, A. T.; Guevarra, K. S.; Hawley, E. L.; Nazaroff, W. W. *Environ. Sci. Technol.* 2002, *36*, 846-853.
[173] Singer, B. C.; Hodgson, A. T.; Nazaroff, W. W. *Atmos. Environ.* 2003, *37*, 5551-5561.
[174] Borbon, A.; Fontaine, H.; Locoge, N.; Veillerot, M.; Galloo, J. C. *Atmos. Environ.* 2003, *37*, 4051-4064.
[175] Borbon, A.; Fontaine, H.; Locoge, N.; Veillerot, M.; Galloo, J. C. *Atmos. Environ.* 2003, *37*, 4065-4076.
[176] Ciccioli, P.; Brancaleoni, E.; Frattoni, M.; Marta, S.; Brachetti, A.; Vitullo, M.; Tirone, G.; Valentini, R. *J. Chromatogr. A* 2003, *985*, 283-296.
[177] Eriksson, K.; Tjärner, D.; Marqvardsen, I.; Järvholm, B. *Chemosphere.* 2003, *50*, 1343-1347.
[178] Guo, H.; Murray, F.; Lee, S. C. *Build. Environ.* 2003, *38*, 1413-1422.
[179] Hodgson, A. T.; Faulkner, D.; Sullivan, D. P.; Dibartolomeo, D. L.; Russell, M. L.; Fisk, W. J. *Atmos. Environ.* 2003, *37*, 5517-5527.
[180] Jo, W. K.; Kim, K. Y.; Park, K. H.; Kim, Y. K.; Lee, H. W.; Park, J. K. *Environ. Res.* 2003, *92*, 166-171.
[181] Komenda, M.; Schaub, A.; Koppmann, R. *J. Chromatogr. A* 2003, *995*, 185-201.
[182] Murena, F.; Vorraro, F. *Atmos. Environ.* 2003, *37*, 4853-4859.
[183] Olsson, M.; Petersson, G. *Sci. Total Environ.* 2003, *303*, 215-220.
[184] Persily, A.; Howard-Reed, C.; Nabinger, S. J. *Atmos. Environ.* 2003, *37*, 5505-5516.
[185] Rabaud, N. E.; Ebeler, S. E.; Ashbaugh, L. L.; Flocchini, R. G. *Atmos. Environ.* 2003, *37*, 933-940.
[186] Saarela, K.; Tirkkonen, T.; Laine-Ylijoki, J.; Jurvelin, J.; Nieuwenhuijsen, M. J.; Jantunen, M. *Atmos. Environ.* 2003, *37*, 5563-5575.
[187] Sorensen, M.; Skov, H.; Autrup, H.; Hertel, O.; Loft, S. *Sci. Total Environ.* 2003, *309*, 69-80.
[188] Takanami, Y.; Chida, M.; Hasebe, H.; Sone, Y.; Suhara, S. *J. Chromatogr. Sci.* 2003, *41*, 317-322.
[189] Wu, C. H.; Lin, M. N.; Feng, C. T.; Yang, K. L.; Lo, Y. S.; Lo, J. G. *J. Chromatogr. A* 2003, *996*, 225-231.
[190] Xie, J.; Wang, X.; Sheng, G.; Bi, X.; Fu, J. *Atmos. Environ.* 2003, *37*, 3365-3374.

[191] Xu, X.; Van Stee, L. L. P.; Williams, J.; Beens, J.; Adahchour, M.; Vreuls, R. J. J.; Brinkman, U. A. T.; Lelieveld, J. *Atmos. Chem. Phys. Discuss.* 2003, *3*, 665-682.

[192] Zou, S. C.; Lee, S. C.; Chan, C. Y.; Ho, K. F.; Wang, X. M.; Chan, L. Y.; Zhang, Z. X. *Chemosphere.* 2003, *51*, 1015-1022.

[193] Zuraimi, M. S.; Tham, K. W.; Sekhar, S. C. *Build. Environ.* 2003, *38*, 23-32.

[194] Baek, S.-O.; Jenkins, R. A. *Atmos. Environ.* 2004, *38*, 6583-6599.

[195] Boy, M.; Petäjä, T.; Dal Maso, M.; Rannik, Ü.; Aalto, P.; Laaksonen, A.; Vaattovaara, A.; Joutsensaari, J.; Hoffmann, T.; Warnke, J.; Apostolaki, M.; Stephanou, E. G.; Tsapakis, M.; Kouvarakis, A.; Pio, C.; Carvalho, A.; Römpp, A.; Moortgat, G.; Spirig, C.; Guenther, A.; Greenberg, J.; Ciccoli, P.; Kulmala, M. *Atmos. Chem. Phys.* 2004, *4*, 657-678.

[196] Fernández-Villarrenaga, V.; López-Mahía, P.; Muniategui-Lorenzo, S.; Prada-Rodríguez, D.; Fernández-Fernández, E.; Tomàs, X. *Sci. Total Environ.* 2004, *334-335*, 167-176.

[197] Guo, H.; Murray, F.; Lee, S. C.; Wilkinson, S. *Build. Environ.* 2004, *39*, 179-187.

[198] Kotzias, D.; Geiss, O.; Leva, P.; Bellintani, A.; Arvanitis, A.; Kephalopoulos, S. *Fresenius Environ. Bull.* 2004, *13*, 1536-1549.

[199] Lai, H. K.; Kendall, M.; Ferrier, H.; Lindup, I.; Alm, S.; Hánninen, O.; Jantunen, M.; Mathys, P.; Colvile, R.; Ashmore, M. R.; Cullinan, P.; Nieuwenhuijsen, M. J. *Atmos. Environ.* 2004, *38*, 6399-6410.

[200] Mukerjee, S.; Norris, G. A.; Smith, L. A.; Noble, C. A.; Neas, L. M.; Özkaynak, A. H.; Gonzales, M. *Environ. Sci. Technol.* 2004, *38*, 2317-2327.

[201] Park, K. H.; Jo, W. K. *Atmos. Environ.* 2004, *38*, 1303-1312.

[202] Sakai, K.; Norbäck, D.; Mi, Y.; Shibata, E.; Kamijima, M.; Yamada, T.; Takeuchi, Y. *Environ. Res.* 2004, *94*, 75-85.

[203] Schlink, U.; Rehwagen, M.; Damm, M.; Ritcher, M.; Borte, M.; Herbarth, O. *Atmos. Environ.* 2004, *38*, 1181-1190.

[204] Simon, V.; Baer, M.; Torres, L.; Olivier, S.; Meybeck, M.; Della Massa, J. P. *Sci. Total Environ.* 2004, *334-335*, 177-183.

[205] Srivastava, A. *Atmos. Environ.* 2004, *38*, 6829-6843.

[206] Tham, K. W.; Zuraimi, M. S.; Sekhar, S. C. *Environ. Int.* 2004, *30*, 1075-1088.

[207] Wang, K. Y.; Shallcross, D. E.; Hadjinicolaou, P.; Giannakopoulos, C. *Water Air Soil Pollut.* 2004, *156*, 29-55.

[208] Wevill, D. J.; Carpenter, L. J. *Analyst.* 2004, *129*, 634-638.

[209] Yamada, E.; Hosokawa, Y.; Matsushita, K.; Fuse, Y. *Anal. Sci.* 2004, *20*, 107-112.

[210] Zhao, L.; Wang, X.; He, Q.; Wang, H.; Sheng, G.; Chan, L. Y.; Fu, J.; Blake, D. R. *Atmos. Environ.* 2004, *38*, 6177-6184.

[211] Back, J.; Hari, P.; Hakola, H.; Juurola, E.; Kulmala, M. *Boreal Environ. Res.* 2005, *10*, 409-424.

[212] Moukhtar, S.; Bessagnet, B.; Rouil, L.; Simon, V. *Atmos. Environ.* 2005, *39*, 3535-3547.

[213] Padhy, P. K.; Varshney, C. K. *Chemosphere.* 2005, *59*, 1643-1653.

[214] Helmig, D.; Ortega, J.; Guenther, A.; Herrick, J. D.; Geron, C. *Atmos. Environ.* 2006, *40*, 4150-4157.

[215] Holzke, C.; Hoffmann, T.; Jaeger, L.; Koppmann, R.; Zimmer, W. *Atmos. Environ.* 2006, *40*, 3174-3185.

[216] Lee, B. S.; Wang, J. L. *Atmos. Environ.* 2006, *40*, 5486-5495.
[217] Noe, S. M.; Ciccioli, P.; Brancaleoni, E.; Loreto, F.; Niinemets, U. *Atmos. Environ.* 2006, *40*, 4649-4662.
[218] Rouvière, A.; Brulfert, G.; Baussand, P.; Chollet, J. P. *Atmos. Environ.* 2006, *40*, 3613-3620.
[219] Zuraimi, M. S.; Roulet, C. A.; Tham, K. W.; Sekhar, S. C.; Cheong, K. W. D.; Wong, N. H.; Lee, K. H. *Build. Environ.* 2006, *41*, 316-329.
[220] Asensio, D.; Peñuelas, J.; Ogaya, R.; Llusia, J. *Atmos. Environ.* 2007, *41*, 2456-2466.
[221] Ribes, A.; Carrera, G.; Gallego, E.; Roca, X.; Berenguer, M. J.; Guardino, X. *J. Chromatogr. A* 2007, *1140*, 44-55.
[222] Bao, H.; Kondo, A.; Kaga, A.; Tada, M.; Sakaguti, K.; Inoue, Y.; Shimoda, Y.; Narumi, D.; Machimura, T. *Environ. Res.* 2008, *106*, 156-169.
[223] Dodson, R. E.; Levy, J. I.; Spengler, J. D.; Shine, J. P.; Bennett, D. H. *Atmos. Environ.* 2008, *42*, 1569-1581.
[224] Su, Y.-C. C., C.-C.; Wang, J.-A. *J. Chromatogr. A* 2008, *1201*, 134-140.
[225] Wang, J. L.; Wang, C. H.; Lai, C. H.; Chang, C. C.; Liu, Y.; Zhang, Y.; Liu, S.; Shao, M. *Atmos. Environ.* 2008, *42*, 6233-6246.
[226] Zhang, Y. H.; Su, H.; Zhong, L. J.; Cheng, Y. F.; Zeng, L. M.; Wang, X. S.; Xiang, Y. R.; Wang, J. L.; Gao, D. F.; Shao, M.; Fan, S. J.; Liu, S. C. *Atmos. Environ.* 2008, *42*, 6203-6218.
[227] Buczynska, A. J.; Krata, A.; Stranger, M.; Locateli Godoi, A. F.; Kontozova-Deutsch, V.; Bencs, L.; Naveau, I.; Roekens, E.; Van Grieken, R. *Atmos. Environ.* 2009, *43*, 311-318.
[228] Chung, K. C.; Stock, T. H.; Smith, L. A.; Afshar, M.; Liao, X. J. L.; Stallings, C. *Environ. Res.* 2009, *109*, 943-951.
[229] Ohura, T.; T., A.; Xueyou, S.; Li, S.; Zhang, P.; Zhu, L. *Atmos. Environ.* 2009, *43*, 6352-6359.
[230] Pérez-Rial, D.; López-Mahia, P.; Muniategui-Lorenzo, S.; Prada-Rodriguez, D. *J. Environ. Monit.* 2009, *11*, 1216-1225.
[231] Winters, A. J.; Adams, M. A.; Bleby, T. M.; Rennenberg, H.; Steigner, D.; Steinbrecher, R.; Kreuzwieser, J. *Atmos. Environ.* 2009, *43*, 3035-3043.
[232] Faubert, P.; Tiiva, P.; Rinnan, A.; Raty, S.; Holopainen, J. K.; Holopainen, T.; Rinnan, R. *Atmos. Environ.* 2010, *44*, 4432-4439.
[233] Hyttinen, M.; Masalin-Weijo, M.; Kalliokoski, P.; Pasanen, P. *Atmos. Environ.* 2010, *44*, 5028-5033.
[234] Jünger, M. B., B.; Baumbach, J. I. *Anal. Bioanal. Chem.* 2010, *396*, 471-482.
[235] Liu, Y.; Shen, J.; Zhu, X.-D. *Environ. Monit. Assess.* 2010, *171*, 249-254.
[236] Pérez-Rial, D.; López-Mahía, P.; Tauler, R. *Atmos. Environ.* 2010, *44*, 5122-5132.
[237] Blanch, J. S.; Llusià, J.; Niinemets, Ü.; Noe, S. M.; Peñuelas, J. *J. Environ. Biol.* 2011, *32*, 1-6.
[238] Joó, É.; Dewulf, J.; Demarcke, M.; Amelynck, C.; Schoon, N.; Müller, J. F.; Simpraga, M.; Steppe, K.; Van Langenhove, H. *Int. J. Mass Spectrom.* 2011, *291*, 90-95.
[239] Pekey, B.; Yilmaz, H. *Microchem. J.* 2011, *97*, 213-219.
[240] Zhou, J. A.; You, Y.; Bai, Z. P.; Hu, Y. D.; Zhang, J. F.; Zhang, N. *Sci. Total Environ.* 2011, *409*, 452-459.
[241] Lai, H.; Corbin, I.; Almirall, J. R. *Anal. Bioanal. Chem.* 2008, *392*, 105-113.

[242] Martin, A. N.; Farquar, G. R.; Jones, A. D.; Frank, M. *Anal. Bioanal. Chem.* 2010, *396*, 739-750.
[243] Prikryl, P.; Sevcik, J. G. K. *J. Chromatogr. A* 2008, *1179*, 24-32.
[244] Wang, A.; Fang, F.; Pawliszyn, J. *J. Chromatogr. A* 2005, *1072*, 127-135.
[245] Berezkin, V. G.; Makarov, E. D.; Stolyarov, B. V. *J. Chromatogr. A* 2003, *985*, 63-65.
[246] Saito, Y.; Ueta, I.; Ogawa, M.; Abe, A.; Yogo, K.; Shirai, S.; Jinno, K. *Anal. Bioanal. Chem.* 2009, *393*, 861-869.
[247] Saito, Y.; Ueta, I.; Kotera, K.; Ogawa, M.; Wada, H.; Jinno, K. *J. Chromatogr. A* 2006, *1106*, 190-195.
[248] Jochmann, M. A.; Yuan, X.; Schilling, B.; Schmidt, T. C. *J. Chromatogr. A* 2008, *1179*, 96-105.
[249] Esteve-Turrillas, F. A.; Pastor, A.; De La Guardia, M. *Anal. Chim. Acta.* 2007, *593*, 108-116.
[250] Morley, M. P., J. *IEEE Sens. J.* 2010, *10*, 167-172.
[251] Delcourt, J.; P., G. J.; Muller, J. *Chromatographia.* 1983, *17*, 88-92.
[252] Esteban, J. L.; Martínez-Castro, I.; Sanz, J. *J. Chromatogr. A* 1993, *657*, 155-164.
[253] Czaplicka, M.; Klejnowski, K. *J. Chromatogr. A* 2002, *976*, 369-376.
[254] Campos-Candel, A.; Llobat-Estelles, M.; Mauri-Aucejo, A. R. *Anal. Bioanal. Chem.* 2007, *387*, 1517-1523.
[255] Richter, B. E.; Jones, B. A.; Ezzell, J. L.; Porter, N. L. *Anal. Chem.* 1996, *68*, 1033-1039.
[256] Zhu, X. L.; Su, Q. D.; Cai, J. B.; Yang, J. *Anal. Chim. Acta.* 2006, *579*, 88-94.
[257] Ramírez, N.; Cuadras, A.; Rovira, E.; Borrull, F.; Marce, R. M. *Talanta.* 2010, *82*, 719-727.
[258] Cao, X.-L.; Hewitt, N. C. *Reactive hydrocarbons in the atmosphere. The sampling and analysis of volatile organic compounds in the atmosphere*; Academic Press: Lancaster, UK, 1999; pp 120-157.
[259] Santos, F. J.; Galceran, M. T. *TrAC Trends Anal. Chem.* 2002, *21*, 672-685.
[260] Dewulf, J.; Van Langenhove, H. *TrAC Trends Anal. Chem.* 2002, *21*, 637-646.
[261] Qi, M. L.; Armstrong, D. W. *Anal. Bioanal. Chem.* 2007, *388*, 889-899.
[262] De Alencastro, L. F.; Grandjean, D.; Tarradellas, J. *Chimia* 2003, *57*, 499-504.
[263] Dunn, M.; Shellie, R.; Morrison, P.; Marriott, P. *J. Chromatogr. A* 2004, *1056*, 163-169.
[264] Mondello, L.; Tranchida, P. Q.; Dugo, P.; Dugo, G. *Mass Spectrom. Rev.* 2008, *27*, 101-124.
[265] Liu, Z. Y.; Phillips, J. B. *J. Chromatogr. Sci.* 1991, *29*, 227-231.
[266] Vendeuvre, C.; Bertoncini, F.; Duval, L.; Duplan, J. L.; Thiebaut, D.; Hennion, M. C. *J. Chromatogr. A* 2004, *1056*, 155-162.
[267] Phillips, J. B.; Xu, J. *J. Chromatogr. A* 1995, *703*, 327-334.
[268] De Geus, H. J.; De Boer, J.; Brinkman, U. A. T. *J. Chromatogr. A* 1997, *767*, 137-151.
[269] Phillips, J. B.; Gaines, R. B.; Blomberg, J.; Van Der Wielen, F. W. M.; Dimandja, J. M.; Green, V.; Granger, J.; Patterson, D.; Racovalis, L.; De Geus, H. J.; De Boer, J.; Haglund, P.; Lipsky, J.; Sinha, V.; Ledford, J. E. L. *J. High Res. Chromatogr.* 1999, *22*, 3-10.
[270] Kinghorn, R. M.; Marriott, P. J. *J. High Res. Chromatogr.* 1998, *21*, 32-38.
[271] Kinghorn, R. M.; Marriott, P. J. *J. High Res. Chromatogr.* 1998, *21*, 620-622.

[272] Beens, J.; Adahchour, M.; Vreuls, R. J. J.; Van Altena, K.; Brinkman, U. A. T. *J. Chromatogr. A* 2001, *919*, 127-132.
[273] Kinghorn, R. M.; Marriott, P. J.; Dawes, P. A. *J. High Res. Chromatogr.* 2000, *23*, 245-252.
[274] Beens, J.; Dallüge, J.; Adahchour, M.; Vreuls, R. J. J.; Brinkman, U. A. T. *J. Microcolumn Sep.* 2001, *13*, 134-140.
[275] Panic, O.; Gorecki, T. *Anal. Bioanal. Chem.* 2006, *386*, 1013-1023.
[276] Bertsch, W. *J. High Res. Chromatogr.* 1999, *22*, 647-665.
[277] Bertsch, W. *J. High Res. Chromatogr.* 2000, *23*, 167-181.
[278] Cortes, H. J. *J. Chromatogr.* 1992, *626*, 3-23.
[279] Cortes, H. J.; Winniford, B.; Luong, J.; Pursch, M. *J. Separ. Sci.* 2009, *32*, 883-904.
[280] Dalluge, J.; Beens, J.; Brinkman, U. A. T. *J. Chromatogr. A* 2003, *1000*, 69-108.
[281] De Geus, H. J.; De Boer, J.; Brinkman, U. A. T. *TrAC Trends Anal. Chem.* 1996, *15*, 168-178.
[282] Panda, S. K.; Andersson, J. T.; Schrader, W. *Anal. Bioanal. Chem.* 2007, *389*, 1329-1339.
[283] Leckrone, K. J.; Hayes, J. M. *Anal. Chem.* 1997, *69*, 911-918.
[284] Lewis, A. C.; Carslaw, N.; Marriott, P. J.; Kinghorn, R. M.; Morrison, P.; Lee, A. L.; Bartle, K. D.; Pilling, M. J. *Nature.* 2000, *405*, 778-781.
[285] Xu, X.; Williams, J.; Plass-Dulmer, C.; Berresheim, H.; Salisbury, G.; Lange, L.; Lelieveld, J. *Atmos. Chem. Phys.* 2003, *3*, 1461-1475.
[286] Rocha, S. M.; Coelho, E.; Zrostlikova, J.; Delgadillo, I.; Coimbra, M. A. *J. Chromatogr. A* 2007, *1161*, 292-299.
[287] Wu, J.; Lu, X.; Tang, W.; Kong, H.; Zhou, S.; Xu, G. *J. Chromatogr. A* 2004, *1034*, 199-205.
[288] Jobson, B. T.; Parrish, D. D.; Goldan, P.; Kuster, W.; Fehsenfeld, F. C.; Blake, D. R.; Blake, N. J.; Niki, H. *J. Geophys. Res. Atmospheres.* 1998, *103*, 13557-13567.
[289] Demeestere, K.; Dewulf, J.; De Roo, K.; De Wispelaere, P.; Van Langenhove, H. *J. Chromatogr. A* 2008, *1186*, 348-357.
[290] Pérez-Pastor, R. M.; García -Alonso, S.; Queijido-Cabezas, A. J. *Environ. Monit. Assess.* 2002, *75*, 33-50.
[291] Hayward, S.; Muncey, R. J.; James, A. E.; Halsall, C. J.; Hewill, C. N. *Atmos. Environ.* 2001, *35*, 4081-4087.
[292] Kim, J.-C. *Atmos. Environ.* 2001, *35*, 3279-3292.
[293] Sartin, J. H.; Halsall, C. J.; Hayward, S.; Hewitt, C. N. *Atmos. Environ.* 2002, *36*, 5311-5321.
[294] Zhihui, W.; Yuhua, B.; Shuyu, Z. *Atmos. Environ.* 2003, *37*, 3771-3782.
[295] Helmig, D.; Bocquet, F.; Pollmann, J.; Revermann, T. *Atmos. Environ.* 2004, *38*, 557-572.
[296] Borrego, C.; Gomes, P.; Barros, N.; Miranda, A. I. *J. Chromatogr. A* 2000, *889*, 271-279.
[297] Song, K.; Ahn, B.; Jung, E.; Lee, Y. I.; Ko, S. *Anal. Chim. Acta.* 2007, *583*, 210-215.
[298] Maeda, T.; Funaki, K.; Yanaguchi, Y.; Ichioka, K. *J. High Res. Chromatogr.* 1998, *21*, 471-474.
[299] Geron, C.; Guenther, A.; Greenberg, J.; Loescher, H. W.; Clark, D.; Baker, B. *Atmos. Environ.* 2002, *36*, 3793-3802.

[300] Urhahn, T.; Ballshmiter, K. *Fresen. J. Anal. Chem.* 2000, *366*, 365-367.
[301] Hills, A. J.; Zimmerman, P. R. *Anal. Chem.* 1990, *62*, 1055-1060.
[302] Rasulov, B.; Copolovici, L.; Laisk, A.; Niinemets, U. *Plant. Physiol.* 2009, *149*, 1609-1618.
[303] Turner, N.; Jones, M.; Grice, K.; Dawson, D.; Ioppolo-Armanios, M.; Fisher, S. J. *Atmos. Environ.* 2006, *40*, 3381-3388.
[304] Rudolph, J.; Czuba, E.; Norman, A. L.; Huang, L. *Atmos. Environ.* 2002, *35*, 1173-1181.
[305] Rudolph, J.; Anderson, R. S.; Czapiewski, K. V.; Czuba, E.; Ernest, D.; Gillespie, T.; Huang, L.; C, R.; Thompson, A. E. *J. Atmos. Chem.* 2003, *44*, 39-45.
[306] Ruzsanyi, V.; Baumbach, J. I.; Sielemann, S.; Litterst, P.; Westhoff, M.; Freitag, L. *J. Chromatogr. A* 2005, *1084*, 145-151.
[307] Pozzi, R.; Bocchini, P.; Pinelli, F.; Galletti, G. C. *J. Environ. Monit.* 2006, *8*, 1219-1226.
[308] Márquez-Sillero, I.; Aguilera-Herrador, E.; Cárdenas, S.; Valcárcel, M. *TrAC Trends Anal. Chem.* 2011, *30*, 677-690.
[309] Kanu, A. B.; Hill, J., H. H. *J. Chromatogr. A* 2008, *1177*, 12-27.
[310] Perl, T.; Bödeker, B.; Jünger, M.; Nolte, J.; Vautz, W. *Anal. Bioanal. Chem.* 2010, *397*, 2385-2394.
[311] Ketola, R. A.; Kiuru, J. T.; Tarkiainen, V.; Kokkonen, J. T.; Rasanen, J.; Kotiaho, T. *Anal. Chim. Acta.* 2006, *562*, 245-251.
[312] Kourtidis, K.; Ziomas, I.; Zerefos, C.; Gousopoulos, A.; Balis, D.; Tzoumaka, P. *Atmos. Environ.* 2000, *34*, 1471-1480.
[313] Zelinger, Z.; Civis, S.; Janour, Z. *Analyst.* 1999, *124*, 1205-1208.
[314] Zelinger, Z.; Strizík, M.; Kubát, P.; Civis, S. *Anal. Chim. Acta.* 2000, *422*, 179-185.
[315] Asensio, D.; Peñuelas, J.; Filella, I.; Llusia, J. *Plant Soil.* 2007, *291*, 249-261.
[316] Filella, I.; Wilkinson, M. J.; Llusià, J.; Hewitt, C. N.; Peñuelas, J. *Physiol. Plantarum.* 2007, *130*, 58-66.
[317] Grabmer, W.; Kreuzwieser, J.; Wisthaler, A.; Cojocariu, C.; Graus, M.; Rennenberg, H.; Steigner, D.; Steinbrecher, R.; Hansel, A. *Atmos. Environ.* 2006, *40*, S128-S137.
[318] Salisbury, G.; Williams, J.; Holzinger, R.; Gros, V.; Mihalopoulos, N.; Vrekoussis, M.; Sarda-Esteve, R.; Berresheim, H.; Von Kuhlmann, R.; Lawrence, M.; Lelieveld, J. *Atmos. Chem. Phys.* 2003, *3*, 925-940.
[319] Simpraga, M.; Verbeeck, H.; Demarcke, M.; Joó, É.; Amelynck, C.; Schoon, N.; Dewulf, J.; Van Langenhove, H.; Heinesch, B.; Aubinet, M.; Müller, J. F.; Steppe, K. *Atmos. Environ.* 2011, *45*, 2922-2928.
[320] Karl, T.; Crutzen, P. J.; Mandl, M.; Staudinger, M.; Guenther, A.; Jordan, A.; Fall, R.; Lindinger, W. *Atmos. Environ.* 2001, *35*, 5287-5300.
[321] Schnitzhofer, R.; Beauchamp, J.; Dunkl, J.; Wisthaler, A.; Weber, A.; Hansel, A. *Atmos. Environ.* 2008, *42*, 1012-1024.
[322] Han, K. H.; Zhang, J. S.; Knudsen, H. N.; Wargocki, P.; Chen, H.; Varshney, P. K.; Guo, B. *Atmos. Environ.* 2011, *45*, 3034-3045.
[323] Steinbacher, M.; Dommen, J.; Ammann, C.; Spirig, C.; Neftel, A.; Prevot, A. S. H. *Int. J. Mass Spectrom.* 2004, *239*, 117-128.
[324] De Gouw, J. W., Carsten. *Mass Spectrom. Rev.* 2007, *26*, 223-257.

In: Air Pollution Emissions
Editors: Daniel G. Vasiliev and Robert A. Kask
ISBN: 978-1-62100-453-0

Chapter 6

SPATIAL MODELING OF AIR POLLUTION BASED ON TRAFFIC EMISSIONS IN URBAN AREAS*

***Lubos Matejicek*[1]*, Zbynek Janour*[2] *and Michal Strizik*[3]**
[1] Institute for Environmental Studies, Charles University in Prague
Benatska 2, 128 01, Prague, Czech Republic
[2] Institute of Thermomechanics, Academy of Sciences
Dolejskova 5, 182 00, Prague, Czech Republic
[3] Laboratory of Risk Research and Management,
Technical University of Ostrava
Lumirova 13, 700 30, Ostrava, Czech Republic

ABSTRACT

Road traffic becomes a dominant source of air pollution in urban areas. The emissions of inorganic compounds and volatile organic compounds caused by motor vehicles have increased trends, particularly in urban areas, which together with other emitted compounds can cause human health problems in a long-time period. Procedures for estimating of road traffic emissions in the USA and in the European region are based on guidelines and recommendations. In order to implement the guidelines, decision-making tools are necessary to provide information in an easy understandable form. Thus, the spatial information system is needed to manage spatio-temporal data, provide analysis, solve numerical models and visualize the results. Generally, spatial data include street networks and monitoring networks. Temporal data are represented by changes of traffic intensity and by time series of measured pollutants. Geographic information systems (GISs) extended by environmental modeling tools offer advance estimates of traffic emissions improved by the digital terrain data and by the wind flow effects. The estimates of air pollution in the surrounded areas can be based on spatial interpolation. In a wide range of spatial interpolations, the deterministic techniques or the geostatistical methods are used in dependence on the input data and the results provided by the

* A version of this chapter also appears in Environmental Modeling with GIS, edited by Lubos Matejicek, published by Nova Science Publishers, Inc. It was submitted for appropriate modifications in an effort to encourage wider dissemination of research.

exploratory spatial data analysis (ESDA). Each spatial interpolation attached to the map layer defines concentration levels at a specified time period. The time series of map layers can demonstrate the variability of the air pollution distribution, which brings a new insight into this research.

In addition to mapping of the urban air pollution in dependence on the traffic intensity and the measured concentrations, numerical modeling and simulation in wind tunnels are presented as important tools for exploration of the dispersion and the transport of pollutants caused by wind flows and other effects. A number of software tools based on Gaussian dispersion principles are mentioned in the framework of US EPA guidelines.

As a case study, spatial modeling of air pollution is focused on the urban area of the city of Prague, because a number of motor vehicles registered on the Prague territory is growing. In addition to the registered vehicles, the specific phenomenon in the central Europe is represented by the abrupt increase in traffic of trucks. In spite of new methods for reduction of emissions from motor vehicles, urban development and successive reconstructions of existing roads cause other emissions. Thus, the street network also becomes a set of line sources of stirred up suspended particulates generated by the passing vehicles. These dust emissions, so-called secondary dust, are estimated and partially validated by measurements of the fraction PM_{10}. As an example, estimates of nitrogen oxides are carried out by spatial interpolation based on data from the surface monitoring network and the DIAL-LIDAR measurements.

Integration of spatio-temporal data together with environmental modeling tools brings new possibilities to compare spatial interpolations created for individual compounds at the high temporal and spatial resolution. In the framework of the GIS, mapping of air pollution and human exposures in streets, at workplace locations and residence addresses can help to reduce traffic emissions by optimization of transport scenarios.

Keywords: Road traffic; air pollution; spatial modeling; GIS

1. Introduction

Transport-related emissions are rising and there is increasing consensus that the growth in motorized vehicles in urban areas is incompatible with health, and greenhouse-gas emissions. Traffic intensity and air pollution adverse health effects and overall environmental degradation. Mostly, transport is the only sector for which current emissions are higher than in the last seasons. In the USA, passenger-car emissions and emissions from trucks are increasing. In Europe, freight transport has more than doubled since 1970. In eastern parts of Europe, the 1990s are characterized by a major decline in rail transport and a concurrent increase in car and truck use. The escalation of urban transport is also significant in Chinese cities in the last decade. Thus, the transport, especially in urban areas, is widely recognized as a significant and increasing source of air pollution.

Many reviews are focused on regional and global impacts of transport emissions of air pollution, but mostly deal with the impact of emissions on local urban air quality, and consider only emissions related to the road transport. In addition to these reviews, some reports based on data from protection agencies (US EPA or European Environmental Agency) cover air and sea transport. There is also an emphasis on assessing possible solutions to environmental problems related to transport. But, too little action is taken on to reduce the environmental impacts. For all that the catalytic convertors are being fitted to every new gasoline-engine car, which is followed by similar developments in diesel emissions, impacts of road transport on urban air quality is still growing.

In spite of that the mode of transport is dependent on the selected area, which can include also ships and aircrafts, the attached case studies are focused on road traffic and its effects on air pollution in urban areas. Road transport is characterized from other sources of air pollution in that the pollutants are released in very close proximity to human receptors. It also reduces the opportunity for the atmosphere to dilute the pollutants. In addition to the close proximity, the roads in the city centre have buildings alongside that shelter the roads and reduce the wind speed at the sources of emissions to that on open roads. But this increase of transport emissions has not significant impact on damage human health because people usually spend larger amounts of time indoors [1]. In case of the buildings that line city streets, the indoor concentrations of pollutants are mainly dependent on outdoor concentrations from road traffic adjacent to windows and doors.

2. Numerical Modeling Focused on Air Pollution

In order to quantify impacts of transport emissions, dispersion of pollutants is studied by using of numerical modeling [2] or physical modeling [3]. The numerical dispersion models deal with simulation of increased concentrations near the major roads, compared to overall background levels of emitted compounds by motor vehicles. Due to the wide range of car classes, many compounds must be taken into account. Mostly, the dispersion modeling is focused on carbon monoxide (CO), nitrogen oxides (NO_x) ozone (O_3), coarse ($PM_{10\text{-}2.5}$), fine ($PM_{2.5}$) and ultrafine ($PM_{0.1}$) particles, black carbon (BC), polycyclic aromatic hydrocarbons (PAHs), and benzene. The numerical models are mostly based on Gaussian dispersion principles. Many state agencies and local authorities, especially in the USA, use the modeling program Industrial Source Complex-Short Term, Version 3 (ISCST3) approved and recommended by the US EPA. The ISCST3 implements a set of Gaussian plume-based models that are used to calculate downwind concentrations from point, line, and area sources. In dependence on a spatial scale of dispersion modeling, the roads are represented by line sources, and the crossroads, parking places, bus stops and other transport service places can be represented by point or area sources. While the ISCST3 is developed as the steady-state air emission modeling system, another dispersion model called CALPUFF represents a generalized non- steady-state air emission modeling system. The US EPA approves to use the CALPUFF as a guideline model for regulatory applications involving long range transport, and for near-field applications with non-steady-state effects. In case of local prediction of air pollution from road transport, many studies focused on the middle spatial scale tasks are based on these environmental modeling systems [4, 5, 6, 7]. The ISCST3 and the CALPUFF are complemented by other computer tools for processing of meteorological data, spatial modeling of buildings, and visualization of simulation results.

While the ISCST3 or the CALPUFF represent modeling guidelance, many dispersion models are directly focused on dispersion modeling of traffic emission sources. For example, the CALINE, based on Gaussian diffusion equations, employs a mixing zone concept for pollutant dispersion over the roadway. The prediction of pollutant concentrations is carried out for receptors located within 500 meters of the roadway. The inputs contain data about emissions sources, meteorology and site geometry. The special options are included for modeling of air quality near the intersections, street canyons and parking facilities. Nowadays, the CALINE together with the ISCST3 and other dispersion models are included in the US EPA model library, guidelines and recommendations.

3. Modeling in Wind Tunnels

In case of physical modeling, the relation between urban three-dimensional morphology and windiness is explored by wind tunnel experiments. Highly idealized city models consisting of blocks of buildings with street canyons are usually used for construction of the physical models. In addition to the physical models based on blocks of buildings, there are many complements for a more realistic simulation in the wind tunnels. Trees with various shapes of crowns are positioned inside the street canyons. Line sources are embedded in the bottom of the streets to release of the tracer gas for simulation of traffic emissions. Traffic-induced turbulence is simulated by rotating belts with thin plates etc. [8]. Physical modeling is used to be complemented by numerical modeling based on Navier-Stokes equations that are extended by tools for simulation of turbulent flows and calculation of boundary conditions. Many case oriented studies oriented on the urban environment are included in the books and papers focused on computational wind engineering [9]. The evaluation of surface flow fields in the urban environment can be extended by vertical flows caused by daytime heating processes in the summer periods and by other phenomena [10]. Thus, the wind tunnel experiments together with numerical simulation can provide the encouraging insight into air flows and offer to predict transport of air pollutants from traffic emission sources.

4. Methods Based on Exploratory Spatial Data Analysis

Due to required huge amounts of data and time consuming numerical calculations, another ways of environmental modeling are often used to predict concentration levels of pollutants in the neighborhood of emission sources. For example, the thematic maps focused on pollutant concentrations over urban areas are mostly based on spatial interpolation that can create the continuous surface of the pollutant concentration, standard errors of predictions and the probabilities that critical values are exceeded. In case of air pollution based on traffic emissions, the inputs into the interpolation tools are represented by estimated levels of concentration at the major emission sources formed by crossroads and road sections or by a surface monitoring network. The spatial interpolation by itself uses sample points at different locations. The sample points include estimates or measurements of selected pollutants.

In order to carry out the proper spatial interpolation, the exploratory spatial data analysis (ESDA) assists to gain a deeper understanding of the pollutant dispersion by creating histograms, Voronoi maps, quantile plots, trend analysis, semivariogram/covariance clouds and crosscovariance clouds. The ESDA is often integrated in the geographic information system (GIS), which allows highlighting the spatial objects on the map layers together with their attributes. Generally, two groups of interpolation techniques are provided by the GIS: deterministic and geostatistical. While the deterministic techniques use mathematical functions for spatial interpolation, geostatistical tools rely on both statistical and mathematical methods and, in addition to creating the surface, can assess the uncertainty of the predictions. The deterministic techniques include Inverse Distance Weighted (IDW) and polynomial interpolations. The geostatistical techniques are based on various kriging models [11]. Implementation of interpolation procedures is usually described in the GIS guidebooks that give more details about setting of the internal parameters [12].

5. Traffic Emission Inventories in Urban Areas

Inventories of atmospheric emissions from transport have been carried out in the framework of research projects and case oriented studies for decades. They are constructed on a range of scales from individual urban areas to countries or continents. The estimates of air pollution are structured in a different way depending on country legislation and on local rules that govern environmental protection. Thus, the dissimilarity between the inventories made in the USA and Europe reflect not only true differences in emissions, but also distinctions related to the way in which the inventory has been produced [13]. In the USA, many guidelines and traffic emission inventories are accessible through the web pages of the US EPA. The American urban areas are larger than the European urban areas that usually include historical centers linked by narrow streets, and the block of flats in the suburbs interconnected by the arterial roads. In the European Union, the emission inventory guidebooks are mostly prepared by the European Environmental Agency. Nearly each year, the updates are published and accessible on the Internet. For example, the ENEP/CORINAIR Emission Inventory Guidebook published in 2007 contains chapters focused on road transport [14]. The guidelines are discussed and applied in a number of studies [15, 16, 17]. Many modeling tools for estimation of traffic emissions in European cities are also based on CORINAIR emission inventories [18, 19, 20].

6. Spatial Estimation of Air Pollution by Laser Scanning

Remote sensing techniques, mostly used for detection and classification of the surface objects [21, 22], are also implemented in the systems for detection of atmospheric pollutants. Information about the pollutants is derived from spectroscopic interaction between radiation and atmospheric matter. The intensity of laser beams is decreased when passing through polluted areas of the atmosphere. In order to obtain the most precise measurements, the DIALs (Differential Absorption LIDARs) are employed in the mobile stations for surface measurements of the pollutants over the city road network. It represents active remote sensing methods where the pulse LIDAR techniques can estimate the distribution of the pollutants hundreds meters over the surface. The laser pulses are emitted in two different wavelengths, resonance and reference. The resonance wavelength is chosen within the absorption band of the measured air pollutant. Thus, the decrease of the intensity is higher than for the intensity of the reference wavelength that is outside of the absorption band. A portion of the laser light of both wavelengths is returned back to the DIAL system by molecules of gases (Rayleigh scattering) and by aerosols (Mie scattering). By comparing signal intensities on both wavelengths, the high resolution detector and processing unit can create the pollutant concentration profile along the trajectory of the laser pulses. The emitor and the detector are managed together by the rotary periscope that emits the laser beams and receives the backscarttered signals from any direction.

Applications of the LIDAR and the LIDAR-DIAL techniques show a growing capacity in environmental monitoring, especially, if spatial and temporal distributions are required [23]. A wide range of atmospheric contaminants, as for example volatile organic compounds, nitrogen oxides, ozone, sulphur dioxide and carbon monoxide, can be detected in the range of a few kilometers [24, 25, 26]. In addition to the measurement of individual pollutants over the urban areas, the horizontal and vertical concentration profiles allow estimating of the mixing height and the relationship between vertical mixing and ground-level pollutant concentrations. Thus, the ground-level concentrations of ozone and nitrogen oxides in urban canyons based on DIALs techniques can be compared with the conventional fixed point sampling of pollution gas concentrations obtained by conventional monitoring networks [27]. The measurements of the vertical and temporal variations based on a few observation periods bring other insight into air pollution transport and chemical interactions during a day period [28]. In order to analyze spatio-temporal measurements of air pollution together with surface spatial objects and traffic sources of pollution, the GISs are often used for the creation of map layers extended by concentration profiles [29]. In addition to the spatial data management and visualization, GISs can remarkably assist in spatial interpolation of the horizontal and vertical profiles derived from DIALs measurements [30].

7. GIS-Based Spatial Analysis and Visualization

GIS environment offers tools for exploratory spatial data analysis that can explore fixed point sampling and DIALs concentration profiles in dependency on surface emission sources, terrain models and other surface objects [31, 32]. Generally, GIS offers to integrate a wide range of spatial data, which can support spatial modeling and in some cases dynamic modeling. Various scenarios of data integration for environmental modeling are described in the book written by Goodchild et al. [33, 34]. In spite of a huge offer of the GIS analytical tools [35, 36, 37], there is still the need to develop new extensions that solve case oriented tasks in the framework of environmental analysis and modeling [38]. In some cases, the file exchange or sharing methods between the GIS and the standalone modeling tools must be used to solve large environmental projects that deal with decision-making processes.

The time series of pollutant concentrations from conventional fixed point sampling networks can be linked to the map layers together with the emission sources. The horizontal and vertical profiles from the DIALs offer to extend standard monitoring techniques by exploring pollutant concentrations over the surface. In case of numerical modeling and wind tunnel studies, the GIS extensions are needed to solve time consuming numerical calculations and to create the interconnection between the GIS and the standalone software tools based on US EPA modeling guidelines and other accessible resources. Nowadays, the RDBMS extended by modules for spatial data management, like ESRI's geodatabase, are used in order to support spatial analysis, numerical modeling and visualization in a more complex way that offers direct integration of spatial data formats, time series and analytical tools [39].

8. A Case Study for Spatio-Temporal Analysis of Air Pollution from Urban Traffic in Prague

In order to inform local authorities responsible for controlling road traffic in Prague, monitoring system of air pollution and other risks has been operating for a decade. This monitoring system was developed in the framework of the project, which is shared by the Department of Informatics at the Prague City Hall, by the Czech Hydrometeorological Institute and by the T-Soft company. Aggregated information about air pollution based on the surface monitoring network is presented on the Internet and accessible to public. The ground-level air pollution assessment shares data archived in the database of the Air Quality Information System of the Czech Republic. As an example, Figure 1 illustrates the temporal changes of nitrogen oxides during the time period of a few hours. The upper part contains buttons for interactive data management (functions: refresh, backward, forward, legend and report). The graph can display emission limit concentrations for emergency and regulation that are complemented by a list of selected monitoring stations (accessible on the right side).

Generally, the emission sources generating air pollutants are classified in the framework of the Air Pollution Source Register into a few classes in dependence on risk assessment. In case of the stationary emission sources, there are three classes in the register:

- extremely large and large source combustion processes with the heat-generating capacity above 5 MW and very significant technologies,
- middle-sized source combustion processes with the heat-generating capacity 0.2–5 MW and significant technologies,
- small source combustion processes with the heat-generating capacity under 0.2 MW and less significant technologies.

Monitoring of stationary emission sources is continuously updated and revalued in the framework of implementing of the new monitoring techniques and modeling procedures. The individual class is represented by the mobile emission sources, which are consecutively categorized in dependence on their shape to the point sources (crossroads, portals and tunnel ventilation exhausts, parking places) and the line sources (sections of the urban roads). In order to estimate the emissions from mobile sources in a more precise way, a few projects supported by the Ministry of the Environment of the Czech Republic have been created in the last decade. As an example, the map layer based on a model calculation of traffic emissions is presented in Figure 2. Model calculations take into account the type of vehicles, the composition of traffic flows, the speed, the slope and the estimates of increased emissions generated at a cold start of the vehicle engine. The following pollutants can be included into the tools: PM_{10}, sulphur dioxide, nitrogen oxides, carbon monoxide and hydrocarbons [40].

The spatial interpolations based on data from the surface monitoring stations represent other source of information about living conditions in urban areas. The map layers focused on nitrogen oxides created in the GIS environment with the IDW and with the ordinary kriging are illustrated in Figure 3.

The distribution of the nitrogen dioxide in the street canyon located in the urban area of the city Prague is illustrated in Figure 4. The 3D scenes based on processing of the LIDAR-DIAL data are focused on the arterial street with heavy traffic. The measurements of the concentration located in spatial nodes on the laser lens are interpolated by the IDW and ordinary kriging. As an example, a few time periods are included into the output presentation.

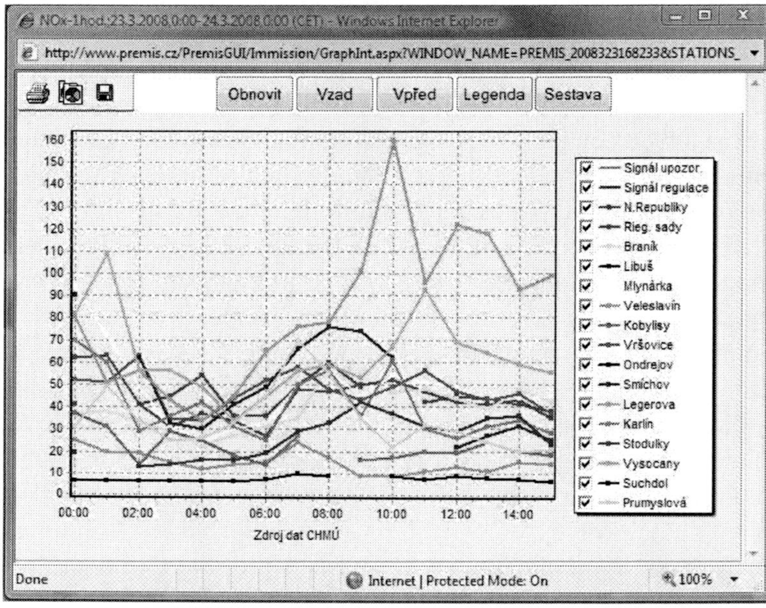

http://www.premis.cz, March-2008.

Figure 1. The actual information about air pollution in Prague accessible for public through the Internet.

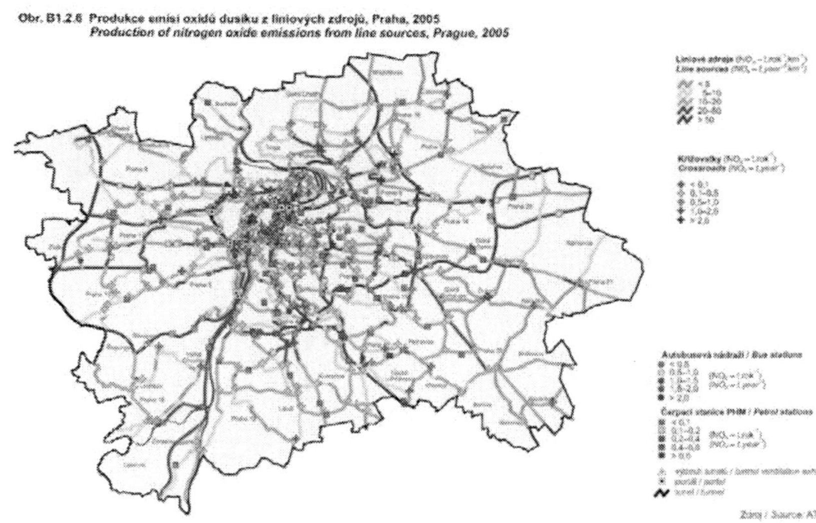

PRAGUE–ENVIRONMENT 2005, MHMP, 2006; http://envis.praha-mesto.cz/, March-2008.

Figure 2. Map layers focused on modeling of nitrogen oxides from line sources.

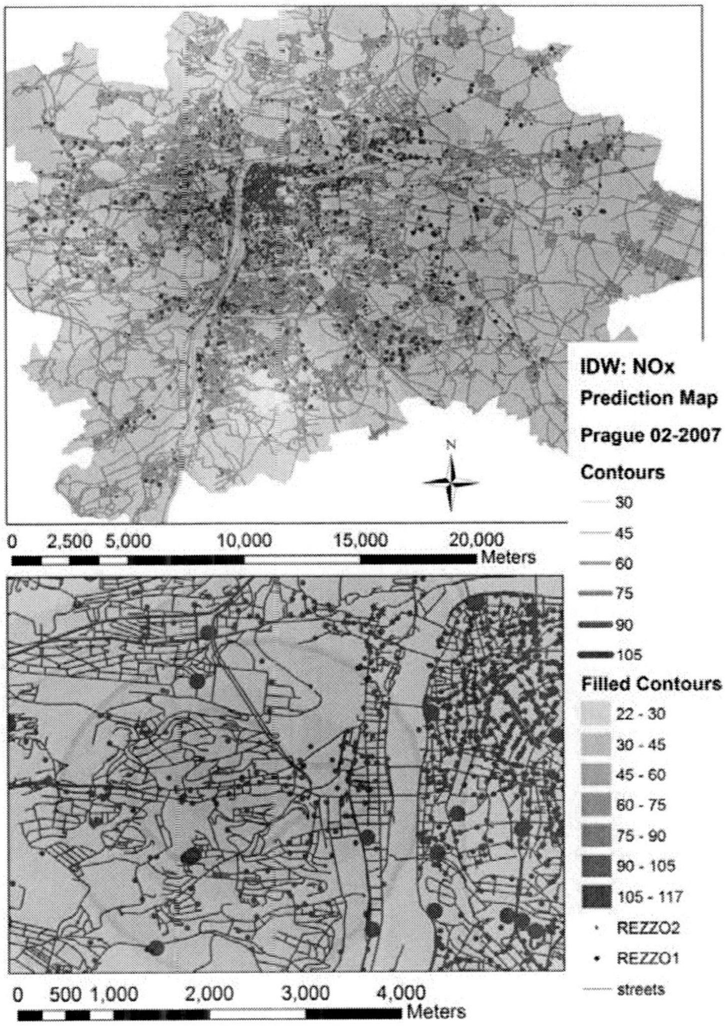

Figure 3. Map layers created in the GIS environment with spatial interpolations (IDW and ordinary kriging) based on data from surface monitoring network, the stationary sources of air emissions are marked in the map layer as points (REZZO1, REZZO2-the Czech guidelines).

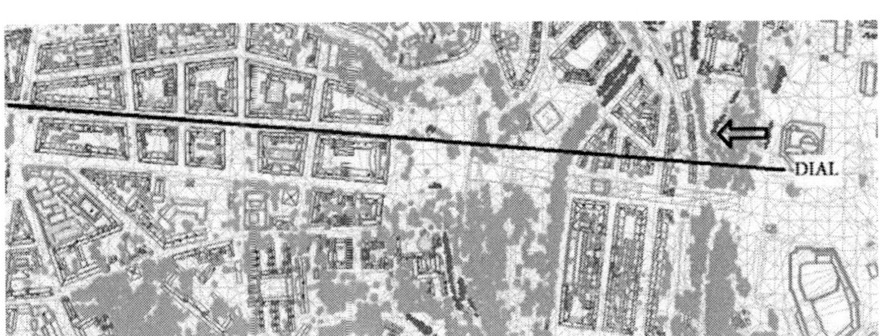

The location of the DIAL's vertical profile in the canyon of an arterial street in Prague.

Figure 4. (Continued).

Visualization of the urban area complemented by the vertical image based on DIAL's data.

The detail view on the street canyon complemented by the image based on DIAL's data.

Figure 4. Processing of the DIAL-LIDAR data in the GIS and CAD environment.

CONCLUSION

A wide range of monitoring techniques and modeling procedures have been mentioned and shortly described in order to present the research activities focused on influence of the urban traffic on living conditions in urban areas. At present, the automotive traffic becomes the most important source of air pollution on the territory of large cities. Thus, the US EPA and the European Environmental Agency carry out many guidelines in the framework of various projects. The projects are shared by state agencies, research associations and universities from the whole world. Consecutively, the research outputs are implemented to the national legislation and guidelines [13, 14].

The environmental modeling supported by powerful computer tools and monitoring systems can predict dispersion and transport of pollutants. The environmental models used for regulatory applications are able to assist in decision-making processes of state authorities and local agencies. In order to improve the precision of predictions, there is still the need to develop new environmental models and to implement new monitoring techniques. A new research is represented by using data from high resolution satellite systems to identify surface objects and their properties in a more precise way. From this point of view, the remote sensing based on the DIAL-LIDARs represents the derived method that can bring a new insight into more detailed on-site air pollution estimates. The relationship between the urban morphology and the air pollution distribution can also help in the validation of numerical models and in a more realistic simulation in the wind tunnels [29, 30].

The attached case study focused on air pollution of the selected urban area in Prague introduces methodology implemented on the local level of the city Prague. In addition to the widely used techniques, spatial interpolations in the GIS environment based on data from the surface monitoring stations and the DIAL-LIDAR measurements are carried out to explore and visualize the distribution of the nitrogen oxides. However, detailed on-site air pollution estimates by the DIAL-LIDAR are not available for each road and time. But, in comparison with the estimates based on surface measurements and the standard modeling procedures, it can improve the prediction of air pollution in higher levels over the surface [41].

ACKNOWLEDGMENTS

The attached case study is partly based on the data created in the framework of the project GACR 205050898, which was generally focused on measurement and modeling of air pollution in urban areas and flat landscape. The authors wish to thank the Lidar company managed by A. Cerny, P. Berger, M. Strizik and P. Engst.

The input spatial data for building of the 3D scenes are originally provided by the Institute of Municipal Informatics of Prague. The map layers of emission sources (registered as REZZO 1 and REZZO 2) and the temporal measurements of the surface air pollution transferred from the Internet are administrated by the IT Department of the Prague City Hall.

The spatio-temporal data were processed by ESRI-ArcGIS and Leica Geosystems-ERDAS Imagine in the GIS Laboratory, Faculty of Natural Science, Charles University in Prague in the framework of the project AVCR 1ET400760405 supported by the Information Society, Academy of Sciences in the Czech Republic.

REFERENCES

[1] Freijer, J.I., Bloemen, H.J.T., Loos, S., Marra, M., Rombout, P.J.A., Steentjes, G.M. and Veen, M.P. (1998). Modelling exposure of the Dutch population to air pollution. *Journal of Hazardous Materials 61*, 107-114.

[2] Venkatram, A., Isakov, V., Thoma, E. and Baldauf, R. (2007). Analysis of air quality data near roadways using a dispersion model. *Atmospheric Environment 41*, 9481-9497.

[3] Skote, M., Sandberg, M., Westerberg, U., Claesson, L. and Johansson A.V. (2005). Numerical and experimental studies of wind environment in an urban morphology. *Atmpospheric Environment 39*, 6147-6158.

[4] Wang, L., Parker, D.B., Parnell, C.B., Lacey, R.E. and Shaw, B.W. (2006). Comparison of CALPUFF and ISCST3 models for predicting downwind odor and source emission rates. *Atmospheric Environment 40*, 4663-4669.

[5] Song, Y., Zhang, M. and Cai, X. (2006). PM10 modeling of Beijing in the winter. *Atmospheric Environment 40*, 4126-4136.

[6] Souto, M.J., Souto, J.A., Munuzuri, V.P., Casares, J.J. and Bermudez, J.L. (2001). A comparision of operational Lagrangian particle and adaptive puff models for plume dispersion forecastting. *Atmospheric Environment 35*, 2349-2360.

[7] Hayes, E.T., Curran, T.P. and Dodd, V.A. (2006). A dispersion modeling approach to determine the odour impact of intensive poultry production units in Ireland. *Bioresource Technology 97*, 1773-1779.

[8] Gromke, C. and Ruck, B. (2007). Influence of trees on the dispersion of pollutants in an urban street canyon-Experimental investigation of the flow and concentration field. *Atmospheric Environment 41*, 3287-3302.

[9] Meroney, R.N. and Bienkiewicz, B. (1997). *Computational Wind Engineering 2.* Amsterdam: Elsevier.

[10] Fujino, T., Asaeda, T. and Ca, V.T. (1999). Numerical analyses of urban thermal environment in a basin climate-application of a k-ε model to complex terrain. *Journal of Wind Engineering and Industrial Aerodynamics 81*, 159-169.

[11] Cressie, N.A.C. (1993). *Statistics for Spatial Data.* New York: Wiley.

[12] Johnston, K., Ver Hoef, J.M., Krivoruchko, K. and Lucas, N. (2001). *Using ArcGIS Geostatistical Analyst.* Redlands, California: ESRI Press.

[13] Seika, M., Metz, N. and Harrison, R.M. (1996). Characteristics of urban and state emission inventories-comparison from Europe and the United States. *The Science of the Total Environment 189/190*, 221-234.

[14] EMEP/CORINAIR Emission Inventory Guidebook-2007. Technical report No 16/2007. Copenhagen: European Environmental Agency.

[15] Salt, J.E. and Moran, A. (1997). International greenhouse gas inventory systems: A comparison between CORINAIR and IPCC methodologies in the EU. *Global Environmental Change 7(4)*, 317-336.

[16] Borrego, C., Tchepel, O., Barros, N. and Miranda, A.I. (2000). Impact of road traffic emissions on air quality of the Lisbon region. *Atmospheric Environment 34*, 4683-4690.

[17] Thorpea, A.J., Harrisona, R.M., Boulterb, P.G. and McCrae, I.S. (2007). Estimation of particle resuspension source strength on a major London Road. *Atmospheric Environment 41*, 8007–8020.

[18] Monforti, F. and Pederzoli, A. (2005). THOSCANE: a tool to detail CORINAIR emission inventories. *Environmental Modelling & Software 20(5)*, 505-508.

[19] Costa, M. and Baldasano, J.M. (1995). Development of a source emission model for atmospheric pollutants in the Barcelona area. *Atmospheric Environment 30(2)*, 309-318.

[20] Klimont, Z., Amann, M., Cofala, J., Gyárfáš, F., Klaassen, G. and Schöpp, W. (1994). An emission inventory for the central European initiative 1988. *Atmospheric Environment 28(2)*, 235-246.

[21] Paine, D.P. and Kiser, J.D. (2003). *Aerial Photography and Image Interpretation.* New York: Wiley.

[22] Lassueur, T., Joost, S. and Randin, C.F. (2006). Very high resolution digital elevation models: Do they improve models of plant species distribution? *Ecological Modelling 198*, 139-153.

[23] Molero, F. and Jaque, F. (1999). The laser as a tool in environmental problems. *Optical Materials 13*, 167-173.

[24] Gondal, M.A. and Mastromarino, J. (2000). Lidar system for remote environmental studies. *Talanta 53*, 147-154.

[25] Dell'Aglio, M., Kholodnykh, A., Lassandro, R. and De Pascale, O. (2002). Development of a Ti: Sapphire DIAL system for pollutant monitoring and meteorological application. *Optics and Lasers in Engineering 37*, 233-244.

[26] Fiorani, L. and Durieux, E. (2001). Comparison among error calculation in differential absorption lidar measurements. *Optics & Laser Technology 33*, 371-377.

[27] Kim, S.W., Yoon, C.S., Won, J.G. and Choi, S.C. (2007). Geound-based remote sensing measurements of aerosol and ozone in an urban area: A case study of mixing height evolution and its effect on ground-level ozone concentrations. *Atmospheric Environment 41*, 7069-7081.

[28] Zwoździak, J., Zwoździak, A., Sówka, I., Ernst, K., Stacewicz, T., Szymański, A., Chudzyński, S., Czyżewski, A., Skubiszak, W. and Stelmaszczyk, K. (2001). Some results on ozone vertical distribution in atmospheric boundary layer from LIDAR and surface measurements over the Kamienczyk Valley, Poland. *Atmospheric Research 58*, 55-70.

[29] Matejicek, L. (2005). Spatial modelling of air pollution in urban areas with GIS: a case study on integrated database development. *Advances in Geosciences 4*, 63-68.

[30] Matejicek, L., Engst, P. and Janour, Z. (2006). A GIS-based approach to spatio-temporal analysis of environmental pollution in urban areas: A case study of Prague's environment extended by LIDAR data. *Ecological Modelling 199*, 261-277.

[31] Briggs, D.J., de Hoogh, C., Gulliver, J., Wills, J., Elliott, P., Kingham, S. and Smallbone, K. (2000). A regression-based method for mapping traffic-related air pollution: application and testing in four contrasting urban environments. *The Science of the Total Environment 253*, 151-167.

[32] Pujadas, M., Plaza, J., Terés, J., Artínano, B. and Millán, M. (2000). Passive remote sensing of nitrogen dioxide as a tool for tracking air pollution in urban areas: the Madrid urban plume, a case of study. *Atmospheric Environment 34*, 3041-3056.

[33] Maidment, D.R. (1996). Environmental Modeling within GIS. In M.F. Goodchild, L.T. Steyaert, B.O. Parks (Ed.), *GIS and Environmental Modeling: Progress and Research Issues* (pp. 315-323). Fort Collins: GIS World, Inc.

[34] Fedra, K. (1996). Distributed Models and Embedded GIS: Integration Strategies and Case Studies. In M.F. Goodchild, L.T. Steyaert, B.O. Parks (Ed.), *GIS and Environmental Modeling: Progress and Research Issues* (pp. 413-417). Fort Collins: GIS World, Inc.

[35] Maguire, D.J., Batty, M., and Goodchild, M.F. (2005). *GIS, Spatial Analysis and Modeling*. Redlands, California: ESRI Press.

[36] Mitchell, A. (2005). *The ESRI Guide to GIS Analysis. Volume 2: Spatial Measurements and Statistics*. Redlands, California: ESRI Press.

[37] Maguire, D.J., Batty, M., and Goodchild, M.F. (2005). *GIS, Spatial Analysis and Modeling*. Redlands, California: ESRI Press.

[38] Grünfeld, K. (2005). Integration spatio-temporal information in environmental monitoring data-a visualization approach applied to moss data. *Science of the Total Environment 347*, 1-20.

[39] MacDonald, A. (2001). *Building a Geodatabase*. Redlands, California: ESRI Press.

[40] Solc, J. and Pospisilova, J. (2007). Prague-Environment 2006. J.I.M. Liberec: Prague.

[41] Matejicek, L. (2006). Spatial estimations of air pollution in urban areas by LIDAR measurements. In: A., Voinov, A.J., Jakeman, A.E., Rizzoli (Ed.), *Proceedings of the*

iEMSs Third Biennial Meeting: "Summit on Environmental Modelling and Software". International Environmental Modelling and Software Society, Burlington, USA, July 2006.

In: Air Pollution Emissions
Editors: Daniel G. Vasiliev and Robert A. Kask ISBN: 978-1-62100-453-0

Chapter 7

USING MONITORING DATA TO EVALUATE THE VARIATIONS OF TRAFFIC-RELATED AIR POLLUTION IN TAIWAN FROM 1994 TO 2006*

Tzu-Yi Pai[1, 2]***, Keisuke Hanaki***[2]***,***
Horng-Guang Leu[3] ***and Shuenn-Chin Chang***[3]
[1]Department of Environmental Engineering and Management, Chaoyang University of Technology, Wufeng, Taichung, 41349, Taiwan, R.O.C.
[2]Department of Urban Engineering, School of Engineering, University of Tokyo, 7-3-1 Hongo, Bunkyo-ku, Tokyo 113-8656, Japan
[3] Environmental Protection Administration, Taipei, 10042, Taiwan, R.O.C.

ABSTRACT

The Air Pollution Control Act (APCA) of Taiwan signed in 1975 prescribes the maximum permissible limits of motor vehicle exhausts as well as the monitoring of air pollution, etc. In this study, the concentrations of air pollutants including SO_2, CO, O_3, PM_{10}, NO_2, and non-methane hydrocarbons (NMHC) from background air pollution monitoring stations (BAQMSs) and traffic area air pollution monitoring stations (TAQMSs) were evaluated to comprehend the variations of traffic-related air pollution in Taiwan from 1994 to 2006. The results indicated that the background concentrations of SO_2 increased from 1994, peaked at 1997 and decreased to 4.31 ppb at 2006. The concentrations of traffic area SO_2 decreased from 16.09 ppb to 7.19 ppb during this period. The background concentrations of CO varied between 0.35 ppm and 0.51 ppm, while the concentrations of traffic area CO decreased from 5.20 ppm to 1.17 ppm from 1994 to 2006. The background concentrations of O_3 increased from 1994, peaked at 2004 (34.10 ppb) and maintained at 33.51 ppb at 2006. The concentrations of traffic area O_3

* A version of this chapter also appears in Encyclopedia of Air Pollution (2 Volume Set), edited by Mark E. Sheffield, published by Nova Science Publishers, Inc. It was submitted for appropriate modifications in an effort to encourage wider dissemination of research.

varied between 18.70 ppb and 25.44 ppb during this period. From 1994 to 2006, the background concentrations of PM_{10} varied between 41.55 $\mu g/m^3$ and 60.73 $\mu g/m^3$, while the concentrations of traffic area PM_{10} decreased from 119.39 $\mu g/m^3$ to 69.62 $\mu g/m^3$. The background concentrations of NO_2 varied between 13.93 ppb and 16.54 ppb, while the concentrations of traffic area NO_2 decreased from 55.85 ppb to 31.68 ppb from 1994 to 2006. The background concentrations of NMHC decreased from 0.61 ppm to 0.11 ppm. The concentrations of traffic area NMHC decreased from 2.52 ppb to 0.72 ppb from 1994 to 2006. From 1994 to 2006, SO_2, CO, PM_{10}, NO_2 and NMHC from TAQMSs decreased by 55.3 %, 77.5 %, 41.7 %, 43.3 % and 71.4 %, respectively. Contrarily, the concentrations of O_3 increased by 17.2 %. The traffic area air quality was improved as the transportation loading increased from 1994. Actually, many efforts including vehicle exhaust emission regulation, fuel and mechanical improvement for cars have been made after the signature of APCA in 1975. But according to the analysis, transportation was the major source of SO_2, CO, PM_{10}, NO_2 and NMHC. For further improving the air quality in Taiwan, TEPA shall adopt stricter regulations for traffic area.

Keywords: Traffic area air pollution monitoring stations, SO_2, CO, O_3, PM_{10}, NO_2, non-methane hydrocarbons (NMHC)

INTRODUCTION

In the past two decades, air pollution has improved in most cities in Western Europe, North American, Japan as well as Taiwan. Air pollution reductions have resulted mainly from greater efficiency and pollution-control technologies in factories, power plants, and other facilities (Cunningham and Cunningham, 2006). Although improvements are also achieved in transportation, the regulation efficiencies of mobile pollution sources are not as significant as those of stationary pollution sources because their mobile, emitted characteristics (Faiz et al., 1995; Fischer et al., 2000; Kingham et al., 2000; Lipfert et al., 2006; Pai et al., 2007).

Air pollution from traffic is one of the main factors considered in the environmental assessment of road establishment. The emission and roadside concentration of those regulated pollutants are potentially harmful to the health or well-being of human, animal or plant life, or to ecological systems. Meanwhile, vehicle emissions may also cause a number of health and nuisance problems (Williams and McCrae, 1995; Künzli et al., 2000; Nicolai, 2002; Vargas, 2003; Monzón and Guerrero, 2004; Wu et al., 2005; Hyder et al., 2006; Pénard-Morand et al., 2006; Finkelstein and Jerrett, 2007).

For examples, carbon monoxide (CO) is one of the important components of air pollution caused by traffic exhaust fumes. CO can cause chronic poisoning which reveals its first symptoms as headaches, blurry vision, difficulty in concentration, and confusion (Atimtay, 2002). The results of recent epidemiologic studies also indicated that ground level ozone (O_3) can exacerbate asthma symptoms even at concentrations lower than 80 ppb (8-h average) (Delfino et al., 1998). The evidences from various studies have shown that micron particle is associated with morbidity and mortality rates particularly due to cardiovascular and respiratory illness (Pekkanen et al., 1997; Zhao et al., 2004; de Kok et al., 2006; Meng and Lu, 2007).

In order to control air pollution, maintain public health and the living environment and improve the quality of life, the Air Pollution Control Act (APCA) of Taiwan was signed in

1975. For controlling the transportation air pollution, Article 16 of Chapter 2 prescribes that the competent authorities at all levels may collect air pollution control fees from stationary and mobile pollution sources that emit air pollutants. For mobile pollution sources, fees shall be collected from the vendor or user based on the types and quantity of air pollutants emitted, or from the vendor or importer based on the type, composition and quantity of fuel.

Article 34 of Chapter 3 prescribes that air pollutants emitted by transportation vehicles shall meet the Vehicular Air Pollutant Emission Standards (VAPES). Article 35 prescribes that the owners of transportation vehicles shall maintain the effective operation of the air pollution control equipment of their vehicle and may not remove or modify this equipment. The type, specifications and labeling of the equipment shall meet the regulations. Article 36 prescribes that the manufacture, import, sale or use of fuel supplied for use in transportation vehicles shall comply with the composition standards and property standards for fuel types. In Article 37, for those in-use motor vehicles that are determined to be unable to meet VAPES for transportation vehicles due to poor design or assembly, the central competent authority shall order the manufacturer or importer to recall for repair within a limited time period. Article 38 prescribes that the manufactured, marketed or imported motor vehicles shall obtain central competent authority issued vehicle model exhaust testing compliance verification before they may apply for license plates. Article 39 prescribes that the central competent authority in conjunction with the Ministry of Transportation and Communications shall determine regulations for the issuance, revocation and cancellation of the vehicle model exhaust testing compliance verification for motor vehicles, and testing and treatment regulations for the air pollutants emissions of transportation vehicles. Article 40 prescribes that the in-use motor vehicles shall undergo regular air pollutant emissions testing. Article 41 prescribes that the competent authorities may perform irregular air pollutant emissions testing or inspections of in-use transportation vehicles at car parks, at airports, at stations, on roadways, in port zones, on water bodies or at other appropriate locations.

In this study, the concentrations of air pollutants including sulfur dioxide (SO_2), CO, O_3, suspended particulates (particles with a diameter of less than 10 micron, PM_{10}), nitrogen dioxide (NO_2) and non-methane hydrocarbons (NMHC) from various air pollution monitoring stations (AQMSs) were evaluated to comprehend the variations of traffic-related air pollution in Taiwan from 1994 to 2006.

METHODS

Area of Study and Measurements

In APCA, Article 8 of Chapter 2 prescribes that the central competent authority may, based on topographical and meteorological conditions, designate single or multiple special municipalities, counties or cities between which it is possible for air pollutants to circulate as total quantity control zones, determine total quantity control plans, and officially announce and implement total quantity controls. In addition, Article 13 of Chapter 2 prescribes that competent authorities at all levels shall select appropriate locations for the installation of air quality monitoring stations and officially publish air quality conditions at regular intervals.

From 1980, the Environmental Protection Administration in the Central Government of Taiwan (TEPA) began installing automatic air quality monitoring stations and by 2006, there were a total of 76 monitoring stations in the entire nation. TEPA concluded that the data collected by these monitoring stations are representational of each individual region and later divided Taiwan into 7 air quality regions (AQRs) according to their geographical and climatic characteristics. These 7 AQRs are Northern, Chu-Miao, Central, Yun-Chia-Nan, Kao-Ping, I-lan and Hwa-Tung AQRs as shown in Figure 1.

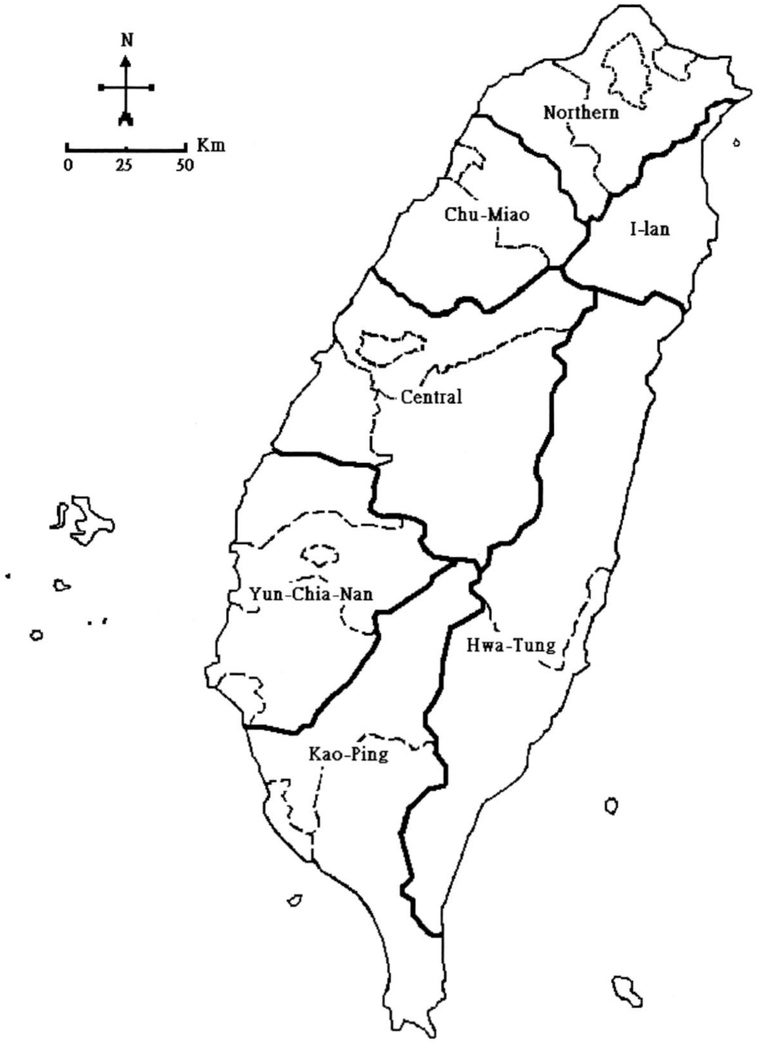

Figure 1. Air quality regions in Taiwan.

Monitoring Stations and Monitoring Items

According to the Article 11 in Air Pollution Control Act Enforcement Rules (APCAER), AQMS designated in Article 13 of APCA shall include five types including general air quality monitoring stations (GAQMS), traffic area air quality monitoring stations (TAQMS), industrial area air quality monitoring stations (IAQMS), national park air quality monitoring stations (PAQMS), and background air quality monitoring stations (BAQMS). The established locations and purposes of different AQMSs are listed in Table 1.

Table 1. The established locations and purposes of different AQMSs

AQMS	Established locations and purposes
GAQMS	established in areas that are densely populated or that may be subject to high pollution or that are able to reflect the air quality distribution in a larger region
TAQMS	established in areas of heavy traffic
IAQMS	established in windy downwind areas industrial parks
PAQMS	established at appropriate sites in national parks
BAQMS	established in areas where there is relatively little human pollution or in windy downwind areas in total quantity control zones

In Article 12 of APCAER, the selection of sites for the air quality monitoring stations in Article 13 of this Act shall take into consideration the following factors including: (1) the type of air quality monitoring stations to be established, (2) the distribution and types of pollution sources and pollutant concentration and distribution, (3) topography, terrain, and meteorological conditions, (4) population distribution and traffic conditions, (5) determination of benefit to control strategy effectiveness, (6) urban plan, regional plan, or other land utilization plan.

The number of air quality monitoring stations established shall conform to the following principles: (1) In accordance with population and habitable area (buildings, paddies, upland fields), one general air quality monitoring station shall be established for every 300,000 persons in areas with a population density exceeding 15,000 persons per square kilometer; one general air quality monitoring station shall be established for every 350,000 persons in areas with a population density below 15,000 persons per square kilometer. The number of air quality monitoring stations may be increased in special municipalities. (2) The number of other types of air quality monitoring stations shall depend on actual needs.

The central competent authority may establish a monitoring center connected with monitoring stations in light of actual needs.

The establishment of air quality monitoring station sampling orifices shall conform to the following principles. First, sampling orifices may not be in locations directly affected by pollution from flues or exhaust outlets, etc. Second, avoid disturbance of air flow and pollutant concentration by nearby obstacles. Third, avoid nearby buildings or obstructing surfaces that may affect pollutant concentration. Forth, determine the height of the sampling orifice above the ground in accordance with the vertical concentration distribution of pollutants near the monitoring station.

Article 13 of APCAER prescribes the pollutant items which shall be tested in different AQMSs. The required test items include SO_2, CO, O_3, PM_{10}, NO_2 and NMHC. The analytical methods for SO_2, CO, O_3, PM_{10}, NO_2 and NMHC are ultraviolet fluorescence method, nondispersive infrared method, ultraviolet absorption method, β-ray attenuation method (or tapered element oscillating microbalance technology), chemiluminescence method and flame ionization detector, respectively.

Results and Discussion

Different Pollutant Variations in TAQMSs from 1994 to 2006

The data were analyzed individually over a long period of time. Figure 2 shows the average values of different pollutants in TAQMSs from 1994 to 2006. According to Figure 2, all pollutant average concentrations declined year by year from 1994 excepting O_3. The concentrations of SO_2, CO, PM_{10}, NO_2 and NMHC decreased to the lowest values of 6.63 ppb, 1.16 ppm, 61.36 $\mu g/m^3$, 31.3 ppb and 0.69 ppm, respectively in 2002, 2004, 1998, 2005 and 2004. From 1994 to 2006, SO_2, CO, PM_{10}, NO_2 and NMHC decreased by 55.3 %, 77.5 %, 41.7 %, 43.3 % and 71.4 %, respectively. Contrarily, the concentrations of O_3 increased from 19.49 ppb to 22.85 ppb and by 17.2 % from 2000 to 2006 (TEPA, 2006).

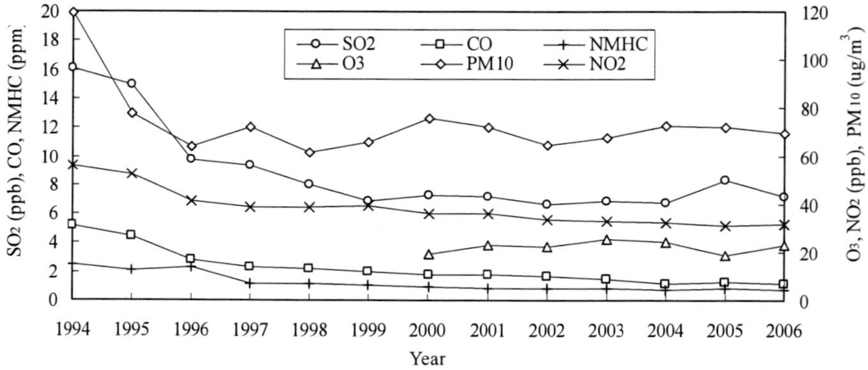

Figure 2. The average values of different pollutants in TAQMSs from 1994 to 2006.

Comparisons of SO_2 Variations between TAQMSs and Other AQMSs

Figure 3 (a) depicts the declining trends of SO_2 between TAQMSs and other AQMSs from 1994 to 2006. According to Figure 3 (a), the concentrations of SO_2 emitted from PAQMSs were the lowest and those from GAQMSs and BAQMSs were the second low. From 1994 to 2000, industries contributed the majority of SO_2, but the SO_2 from transportation was higher than those from industries. In 2000, SO_2 of TAQMSs has exceeded by 1.49 ppb.

From 1994 to 2006, the reduction percentages of SO_2 for GAQMSs, TAQMSs, IAQMSs, PAQMSs and BAQMSs were 43.0 %, 55.3 %, 68.3 %, 40.4 % and 27.3 %, respectively. Although a high level reduction of 55.3 % for TAQMSs was reached, the reduction of 68.3 % from IAQMSs was even higher.

Table 2 lists the correlation coefficients of SO_2 between the TAQMSs and other AQMSs. The SO_2 from TAQMSs were highly correlated with those from other AQMSs ($R > 0.7$). It revealed that the behaviors of SO_2 from different sections would affect each others.

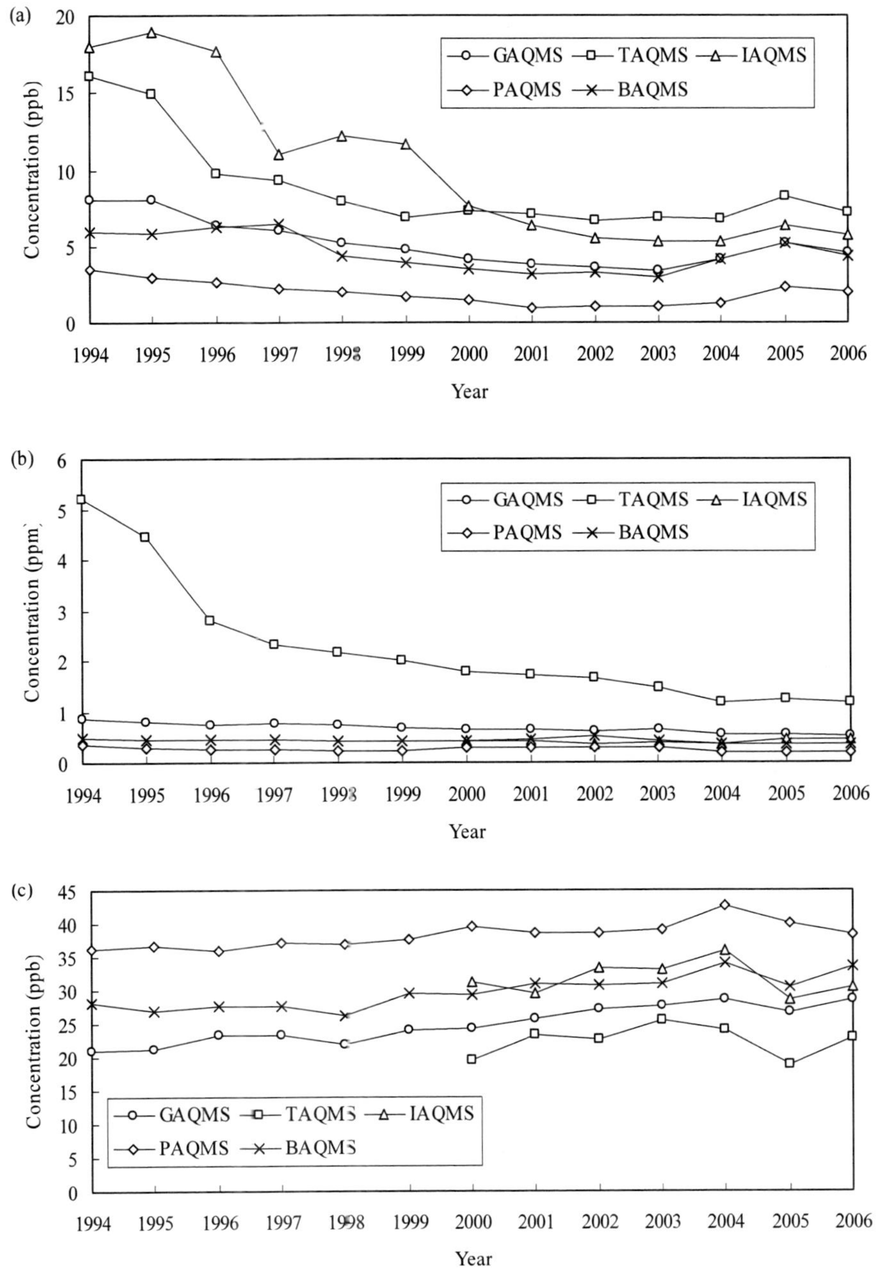

Figure 3. (Continued).

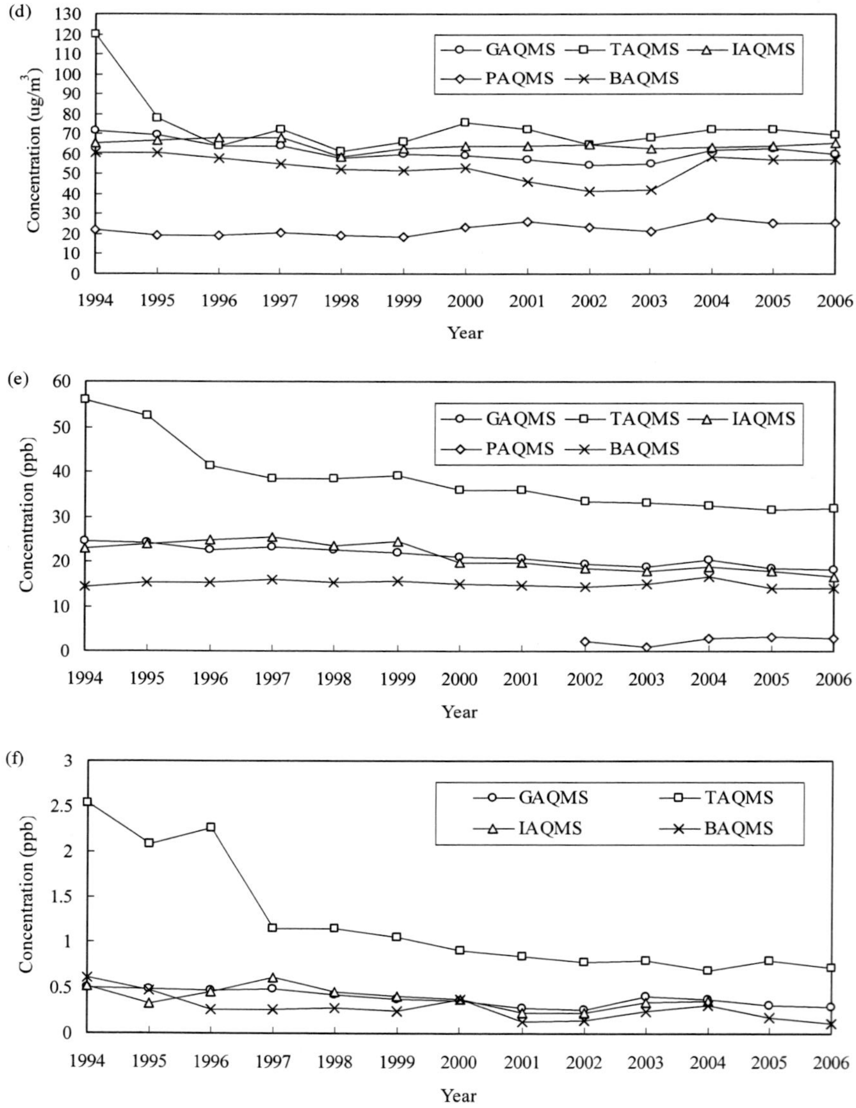

Figure 3. The variations of different pollutants between TAQMSs and other AQMSs from 1994 to 2006. (a) SO_2, (b) CO, (c) O_3, (d) PM_{10}, (e) NO_2 and (f) NMHC.

Table 2. The correlation coefficients between the TAQMSs and other AQMSs

Station Pollutants	GAQMS	IAQMS	PAQMS	BAQMS
SO_2	0.94	0.83	0.87	0.70
CO	0.90	-0.18	0.67	0.61
O_3	0.58	0.60	-0.01	0.51
PM_{10}	0.71	0.24	0.08	0.44
NO_2	0.87	0.65	-0.63	0.03
NMHC	0.78	0.42	-	0.75

Comparisons of CO Variations between TAQMSs and Other AQMSs

The variations of CO between TAQMSs and other AQMSs from 1994 to 2006 are shown in Figure 3 (b). In Figure 3 (b), all CO concentrations were lower than 1.00 ppm excepting those from TAQMSs. Even in 2000, the concentration of CO was still higher than 1.00 ppm.

From 1994 to 2006, the reduction percentages of CO for GAQMSs, TAQMSs, PAQMSs and BAQMSs were 40.2 %, 77.5 %, 44.1 % and 28.6 %, respectively. But the CO concentrations from IAQMSs were contrary to the declining trends of other AQMSs, the values increased by 10.0 % from 2000 to 2006.

When analyzing the correlation coefficients of CO between the TAQMSs and other AQMSs, the CO from TAQMSs were highly correlated with those from GAQMSs, i.e. $R = 0.90$, but the CO from TAQMSs were moderately correlated with those from PAQMSs and BAQMSs ($0.4 < R < 0.7$). The relationship between the CO from TAQMSs and those of IAQMSs revealed a low negative correlation, $R = -0.18$. It revealed that the reductions of CO from different sections were not completely consistent, additionally, the concentrations of CO from TAQMSs would affect those from GAQMSs.

Comparisons of O_3 Variations between TAQMSs and Other AQMSs

When comparing with other pollutants, O_3 revealed an opposite trends. The O_3 concentrations from all AQMSs increased excepting those from IAQMSs. According to Figure 3 (c), the concentrations of O_3 emitted from TAQMSs were the lowest.

From 1994 to 2006, the O_3 concentrations for GAQMSs, PAQMSs and BAQMSs increased by 36.9 %, 5.9 % and 19.4 % respectively. The O_3 concentrations from TAQMSs increased by 17.2 % from 2000 to 2006, contrarily, those from IAQMSs decreased by 2.4 %.

The O_3 from TAQMSs were moderately correlated with those from GAQMSs, PAQMSs and BAQMSs ($0.4 < R < 0.7$).

But the relationship between the O_3 of TAQMSs and those of IAQMSs revealed a very low negative correlation, i.e. $R = -0.01$. It revealed that the increasing behaviors of O_3 from different sections were not completely consistent.

Comparisons of PM_{10} Variations between TAQMSs and Other AQMSs

According to Figure 3 (d), the concentrations of PM_{10} from PAQMSs were the lowest and those from BAQMSs were the second low. From 1994 to 2000, transportation section contributed the majority of PM_{10}, the industries contributed the second quantity.

From 1994 to 2006, the reduction percentages of PM_{10} for GAQMSs, TAQMSs and BAQMSs were 16.9 %, 41.7 % and 6.3 %, respectively. The values from IAQMSs varied between 58.34 ppb and 68.22 ppb, without significant reduction. The reduction was not high from PAQMSs, their values varied between 18.83 ppb and 28.13 ppb.

The PM_{10} from TAQMSs were highly correlated with those from GAQMSs ($R > 0.7$), but the R values between TAQMSs and other three AQMSs (IAQMSs, PAQMSs and BAQMSs) were low. It indicated that the concentrations of PM_{10} from TAQMSs would affect those from GAQMS.

Comparisons of NO_2 Variations between TAQMSs and Other AQMSs

The variations of NO_2 between TAQMSs and other AQMSs from 1994 to 2006 are shown in Figure 3 (e). In Figure 3 (e), all NO_2 concentrations were lower than 30.00 ppb excepting those from TAQMSs. Even in 2000, the concentration of NO_2 was still higher than 30.00 ppb.

From 1994 to 2006, the reduction percentages of NO_2 for GAQMSs, TAQMSs and IAQMSs were 25.7 %, 43.3 % and 27.9 %, respectively. But the NO_2 concentrations from PAQMSs and BAQMSs varied between the ranges of 0.92 – 3.02 ppb and 13.93 - 16.54 ppb, respectively.

When analyzing the correlation coefficients of NO_2 between the TAQMSs and other AQMSs, the NO_2 from TAQMSs were highly correlated with those from GAQMSs, i.e. $R = 0.87$, but the NO_2 from TAQMSs were moderately and lowly correlated with those from IAQMSs and BAQMSs. The relationship between the NO_2 from TAQMSs and those of PAQMSs revealed a moderately negative correlation, $R = -0.63$. It suggested that the concentrations of NO_2 from GAQMSs and TAQMSs affected each others.

Comparisons of NMHC Variations between TAQMSs and Other AQMSs

Figure 3 (f) reveals the variations of NMHC between TAQMSs and other AQMSs. It was noticed that the data from PAQMSs were not available since no detector was installed. All NMHC concentrations were lower than 1.00 ppb excepting those from TAQMSs. Until 2000, the concentration of NMHC began below 1.00 ppb.

From 1994 to 2006, the reduction percentages of NMHC for GAQMSs, TAQMSs, IAQMSs and BAQMSs were 44.0 %, 71.4 %, 100.0 % and 82.0 %, respectively.

The NMHC from TAQMSs were highly correlated with those from GAQMSs and BAQMSs, i.e. $R = 0.78$ and 0.75, but the NMHC from TAQMSs were moderately correlated with those from IAQMSs ($0.4 < R < 0.7$). It revealed that the concentrations of NMHC from TAQMSs would affect those from GAQMSs and BAQMSs.

Regulations to Control the Transportation Air Pollution

According to the analysis, the transportation was the major source of SO_2, CO, PM_{10}, NO_2 and NMHC. According to the correlation analysis, the transportation would affect the concentrations of SO_2, CO, PM_{10}, NO_2 and NMHC in GAQMSs. The concentrations of SO_2 from IAQMSs and PAQMSs would be affected by transportation. The transportation would affect the concentrations of SO_2 and NMHC in BAQMSs.

To control the transportation air pollution, TEPA determines many regulations including Air Pollution Control Fee Collection Regulations, Gas Station Gasoline Vapor Recovery Facility Management Regulations, In-Use Motor Vehicle Recall and Correction Regulations, Regulations Governing Issuance, Revocation, and Cancellation of Compliance Certification for Diesel and Alternative Clean Fuel Engine Vehicle Emissions Inspections, Regulations Governing Issuance, Revocation, and Cancellation of Compliance Certification for Gasoline and Alternative Clean Fuel Engine Vehicle Emissions Inspections, Regulations Governing

Verification Issuance and Cancellation of Certification Compliance for Motorized Bicycle Configuration Emissions, Regulations Governing the Certification and Authorization of Imported Motor Vehicle Air Pollutants and VAPES. Because these regulations, SO_2, CO, PM_{10}, NO_2 and NMHC decreased by 55.3 %, 77.5 %, 41.7 %, 43.3 % and 71.4 %, respectively from 1994 to 2006. But the transportation was still the major source of SO_2, CO, PM_{10}, NO_2 and NMHC. For further improving the air quality in Taiwan, TEPA shall adopt stricter regulations.

CONCLUSION

In order to control air pollution, maintain public health and the living environment and improve the quality of life, the APCA of Taiwan was signed in 1975.

After the signature of APCA, all air pollutant average concentrations declined year by year from 1994 excepting O_3. The concentrations of SO_2, CO, PM_{10}, NO_2 and NMHC decreased to the lowest values of 6.63 ppb, 1.16 ppm, 61.36 $\mu g/m^3$, 31.3 ppb and 0.69 ppm, respectively in 2002, 2004, 1998, 2005 and 2004. From 1994 to 2006, SO_2, CO, PM_{10}, NO_2 and NMHC decreased by 55.3 %, 77.5 %, 41.7 %, 43.3 % and 71.4 %, respectively. Contrarily, the concentrations of O_3 increased from 19.49 ppb to 22.85 ppb and by 17.2 % from 2000 to 2006. But according to the analysis, the transportation was still the major source of SO_2, CO, PM_{10}, NO_2 and NMHC. The correlation analysis showed that the transportation would affect the concentrations of SO_2, CO, PM_{10}, NO_2 and NMHC in GAQMSs. The concentrations of SO_2 from IAQMSs and PAQMSs would be affected by transportation. The transportation would affect the concentrations of SO_2 and NMHC in BAQMSs. For further improving the air quality in Taiwan, TEPA shall adopt stricter regulations for traffic area.

REFERENCES

Atimtay A.T., Emri S., Bagci T., Demir A.U. (2000). Urban CO exposure and its health effects on traffic policemen in Ankara. *Environmental Research*, 82 (3), 222-230.

Cunningham W.P., Cunningham M.A. (2006). *Principles of Environmental Science: Inquiry and Applications.* New York: McGraw-Hill Company.

De Kok, T.M.C.M., Driece H A.L., Hogervorst J.G..F., Briedé, J.J. (2006). Toxicological assessment of ambient and traffic-related particulate matter: A review of recent studies. *Mutation Research-Reviews in Mutation Research*, 613 (2-3), 103-122.

Delfino R.J., Murphy-Moulton A.M., Becklake M.R. (1998). Emergency room visits for respiratory illnesses among the elderly in Montreal: association with low level ozone exposure. *Environmental Research* 76, 67–77.

Faiz A., Gautam S., Burki E. (1995). Air pollution from motor vehicles: issues and options for Latin American countries. *The Science of the Total Environment*, 169 (1-3), 303-310.

Finkelstein M.M., Jerrett M. (2007). A study of the relationships between Parkinson's disease and markers of traffic-derived and environmental manganese air pollution in two Canadian cities. *Environmental Research*, 104 (3), 420-432.

Fischer P.H., Hoek G., van Reeuwijk H., Briggs D.J., Lebret E., van Wijnen, J.H., Kingham S. (2000). Traffic-related differences in outdoor and indoor concentrations of particles and volatile organic compounds in Amsterdam. *Atmospheric Environment*, 34 (22), 3713-3722.

Hyder A.A., Ghaffar A.A., Sugerman D.E., Masood T.I., Ali L. (2006). Health and road transport in Pakistan. *Public Health*, 120 (2), 132-141.

Kingham S., Briggs D., Elliott P., Fischer P., Erik L. (2000). Spatial variations in the concentrations of traffic-related pollutants in indoor and outdoor air in Huddersfield, England. *Atmospheric Environment*, 34 (6), 905-916.

Künzli N., Kaiser R., Medina S., Studnicka M., Chanel O., Filliger P., Herry M., Horak F., Puybonnieux-Texier V. (2000). Public-health impact of outdoor and traffic-related air pollution: a European assessment. *The Lancet*, 356 (9232), 795-801.

Lipfert F.W., Wyzga R.E., Baty J.D., Miller J.P. (2006). Traffic density as a surrogate measure of environmental exposures in studies of air pollution health effects: long-term mortality in a cohort of US veterans. *Atmospheric Environment*, 40 (1), 154-169.

Meng Z., Lu B. (2007). Dust events as a risk factor for daily hospitalization for respiratory and cardiovascular diseases in Minqin, China. *Atmospheric Environment*, 41 (33), 7048-7058.

Monzón A., Guerrero M.J. (2004). Valuation of social and health effects of transport-related air pollution in Madrid (Spain). *Science of the Total Environment*, 334-335, 427-434.

Nicolai T. (2002). Pollution, environmental factors and childhood respiratory allergic disease. *Toxicology*, 181-182, 317 – 321.

Pai T.Y., Hanaki K., Ho H.H., Hsieh C.M. (2007). Using grey system theory to evaluate transportation on air quality trends in Japan, Transportation Research Part D: *Transport and Environment*, 12 (3), 158-166.

Pekkanen J., Timonen K.L., Ruuskanen J., Reponen A., Mirme A. (1997). Effects of Ultrafine and Fine Particles in Urban Air on Peak Expiratory Flow among Children with Asthmatic Symptoms. *Environmental Research*, 74 (1), 24-33.

Pénard-Morand C., Schillinger C., Armengaud A., Debotte G. (2006). Assessment of schoolchildren's exposure to traffic-related air pollution in the French Six Cities Study using a dispersion model. *Atmospheric Environment*, 40 (13), 2274-2287.

Taiwan Environmental Protection Administration (2008). *The Annual Assessment Report of the Air Pollution Control in Taiwan for 2006.* Taipei: Taiwan Environmental Protection Administration. (In Chinese).

Vargas V.M.F. (2003). Mutagenic activity as a parameter to assess ambient air quality for protection of the environment and human health. *Mutation Research/Reviews in Mutation Research,* 544 (2-3), 313-319.

Williams I.D., McCrae I.S. (1995). Road traffic nuisance in residential and commercial areas. *The Science of the Total Environment*, 169 (1-3), 75-82.

Wu J., Lurmann F., Winer A., Lu R., Turco R., Funk T. (2005). Development of an individual exposure model for application to the Southern California children's health study. *Atmospheric Environment*, 39 (2), 259-273.

Zhao L., Wang X., He Q., Wang H., Sheng G., Chan L.Y., Fu J., Blake D.R. (2004). Exposure to hazardous volatile organic compounds, PM_{10} and CO while walking along streets in urban Guangzhou, China. *Atmospheric Environment*, 38 (36), 6177-6184.

INDEX

A

D

E

F

G

J

K

L

M

N

O

P

Q

R

S

W

X

Y

Z